SpringerBriefs in Applied Sciences and Technology

More information about this series at http://www.springer.com/series/8884

Daniel A. James · Nicola Petrone

Sensors and Wearable Technologies in Sport

Technologies, Trends and Approaches for Implementation

Springer

Daniel A. James
SABEL Labs
Griffith University
Nathan, Brisbane, QLD
Australia

Nicola Petrone
Department of Industrial Engineering
Padova
Italy

ISSN 2191-530X ISSN 2191-5318 (electronic)
SpringerBriefs in Applied Sciences and Technology
ISBN 978-981-10-0991-4 ISBN 978-981-10-0992-1 (eBook)
DOI 10.1007/978-981-10-0992-1

Library of Congress Control Number: 2016941605

This text has been peer reviewed using processes administered by the publisher. Reviews were conducted by expert referees to the professional and scientific standards expected of journals published by Springer.

Printed on acid-free paper

This Springer imprint is published by Springer Nature
The registered company is Springer Science+Business Media Singapore Pte Ltd.

Contents

1 Introduction 1
1.1 Overview, Limitations and Approach 2

2 Review 5

3 Sensors 7
3.1 Load and Pressure Measurement 7
3.1.1 Engineering Background 8
3.1.2 Sport Applications 13
3.2 Inertial Sensors 16
3.2.1 Engineering Background 17
3.2.2 Sport Applications 18
3.3 Optical and Other Sensors 20
3.3.1 Engineering Background 20
3.3.2 Sport Applications 20
3.4 Angle and Displacement Sensors 21
3.4.1 Engineering Background 21
3.4.2 Sport Applications 22
3.5 Garment and Apparel 23
3.5.1 Sport Applications 23

4 Approaches 25

5 Implementation 27
5.1 Sensor Selection and Characteristics 27
5.2 Signal Conditioning 30
5.3 Power 30
5.4 Data Acquisition and Memory 31
5.5 Wireless 32

5.6 Data Processing. 33
5.7 Feedback 34
5.8 Packaging. 34

6 Future Directions. 37

7 Conclusions. 39

References 41

Abstract

Engineering practice and the development of new technologies continue to transform many aspects of our lives. The adoption of technology into sport is no exception where advances in materials, equipment design and clothing have had a profound influence on sport. More recently the widespread use of sensors has had an impact on equipment design, together with the practice and performance of sport.

The use of sensors is attractive; they offer the possibility to provide new information in a laboratory context, which was difficult to obtain previously, such as in the ambulatory or 'in field' environment to build a better sporting equipment or a better athlete. The list of application is long and growing rapidly, particularly in the adoption of inertial sensors, force transduction and wearable technologies. Today, micro-sensors have been adopted by mainstream companies as consumer and specialist sport technologies for diverse markets such as gaming, telecommunications and sports. The range of sports where micro-sensors have been used is growing; sensors have been used for purposes such as the improvement of equipment design, the evolution of sport and the monitoring of performance of athletes. Sensor technologies have been implemented in a wide range of sports to provide a broad range of information including load from strain gauges in the design and validation of better skis, motion from inertial sensors in half-pipe snowboarding to detect air time and activity, rower biomechanics and boat movement through the water, force-plate simulations, contact time, step rate and other biomechanical information in running, as well as estimates of energy expenditure.

This paper reviews the adoption of sensors in sport, together with guidelines for the implementation and the identification of likely future trends including wearable technology.

Keywords Sensors · Sports · Force · Pressure · Inertial sensors · Optical sensors · Wearable technologies · Implementation

Chapter 1
Introduction

The scientific study of sport has led to a drive to build better, faster and stronger equipment and ultimately athletes as well. This has largely been driven by the rise of professional sport (and their associated budgets) together with national prestige in international amateur competitions such as the Olympics. Beyond the scope of traditional sciences, advancing technology is seen as a leading edge in sports performance: this has involved a tremendous growth in the adoption of sensors over the past decade, built firmly on the foundation of engineering principles in sport.

The relatively new discipline of Sports Engineering, was once mainly the province of materials and mechanical engineers, but has now come to also include, through the rise of technology, the electronics and computer engineering disciplines. This paper introduces the topic of sensors and instrumentation in sport through a review of these technologies and their current use in sport. This paper will also provide approaches to the use of sensors in sport and implementation examples and guidelines as an aid to those embarking on the use of sensors for the first time or for skilled researchers looking for new horizons.

Athletic and clinical testing for performance analysis and enhancement has traditionally been performed in the laboratory where the required instrumentation is available and environmental conditions can be controlled. In this environment dynamic characteristics of athletes are assessed using indoor treadmills, rowing and cycling machines and even flumes for swimmers or full scale outdoor treadmills for skiing [1]. In general these machines allow for the monitoring of athletes using instrumentation that cannot be used in the training environment. These analysis techniques often require the athlete to remain quasi static thus enabling a constant field of view for optical devices and relatively constant proximity for equipment such as tethered electronic sensors and respiratory gas analysis equipment. The trade-off here, at least in the developmental stages, has been between the accuracy of these very high precision (but often bulky and expensive) systems and the portability of small, light, fast and wearable sensors (of increasing accuracy however) that are demanded by the sport market.

D.A. James and N. Petrone, *Sensors and Wearable Technologies in Sport*,
SpringerBriefs in Applied Sciences and Technology,
DOI 10.1007/978-981-10-0992-1_1

Today by taking advantage of the progress in microelectronics and other micro-technologies [2] it is possible to build instrumentation that is small and light weight enough to be unobtrusive for a number of sporting and clinical applications [3]. Sports fields have the potential to become a sports laboratory: users wearing sensing technologies will either be able to have an immediate feed back during their activity or to access, retrieve, process, share later on the data that were sent to a surrounding data cloud. Researchers as well will benefit of these new opportunities, as the number of functional data available for statistical product analysis and developed will increase enormously.

1.1 Overview, Limitations and Approach

Most phenomena of interest to the study of sport where measurement and quantification are desired are not immediately available or electrically based. The conversion to an electrical signal of any physical event is desirable because it can be manipulated and stored [4]. The use and application of sensors that can transduce a physical event to electrical signals in a sporting context will be the primary focus of this work. We also consider the human body itself, which provides a number of electrical sources that can be non-invasively detected at or near the skin's surface.

Sensors for the measurement of the body and its associated electrical signals is not a new idea and is widely and routinely used in the health and sports sciences disciplines. The focus of this review however, is on the more recent use of sensors in sport with particular focus on technologies still in the research or near commercial stages. Existing commonplace technologies will also be discussed, particularly where the application is suitably novel, for example the use of force plates in climbing holds [5]. Similarly, the definition of sensors itself resists the classification of several technologies that could be considered sensors in the true sense of the word, yet would seem to beyond the scope of such are given a light treatment in the text. These include the widespread adoption of video, particularly its more recent use as both a visual and distancing tool [6, 7] and Global Positing Systems (GPS) [8]. These popular technologies are worthy topics in and of themselves and often used as validation and evaluation methods for sensor development [9].

This review limits itself primarily to developmental applications of sensors in sporting endeavours, though their use in sporting related disciplines is widespread. Whilst sensor adoption in sport owes much of its historical evolution to the medical disciplines, where there is often considerable cross over in ambulatory monitoring, there is also the widespread use of sensors in gaming technologies where they are used for user input and to create a more natural immersive environment for consumers [10]. Although these and other disciplines have the potential to intersect with sport, they are not the primary subject of this review.

The approach of this review is largely systematic, using a three pass method. Initially it uses keyword review of sensors by publication frequency in a leading conference series, before proceeding to a targeted literature review of applications of the sensors and further discovery from those papers. As Sports Engineering is still considered an emerging field, ‘google scholar’ was used primarily, rather than the more established indexes, which are generally limited to the more traditional disciplines, which don’t reflect the usage of sensors in sport as a primary focus.

In the following section the review examples from the literature are drawn on to present guidelines to the use of sensors and their accompanying instrumentation and computational support systems. This section serves as an overview to guide the approach to using sensors, due consideration for limitations and possibilities, together with signposts to more detailed works and likely and emerging future trends.

Chapter 2
Review

The initial keyword review used the conference publication "The Engineering of Sport" for 2012 [11]. This fast turn-around publication serves as a 'mark in the sand' of research works that is both early and timely, as a window into the use of sensors. A review of the publication from a decade earlier in 2002 [12] serves to indicate a sense of emerging trends in sensors and their growing popularity. Individual papers were assessed on their use of sensors and are grouped accordingly in the Table 2.1. This broad-brush stroke shows the progressive adoption of sensors into the Sports Engineering discipline showing growth from 42–71 % of paper volume over the last decade. Of the summarised topics the measures of force (direct and indirect), use of sensors in wind tunnels and inertial sensors were the highest areas of paper frequency and will largely be the focus of the review section. We will largely exclude the routine use of video and motion capture systems, with some exceptions.

The primary areas of focus of these sensor related papers included from 2002 were materials, impacts, vibration, fluid dynamics, optimisation, robotics, measurement, motion analysis and biomechanics. In 2012 the primary areas were aerodynamics, biomechanics, footwear, innovation and design, measurement and instrumentation, modeling and simulation, motion analysis, and sports surfaces. Sensor technologies included human electrodes, resistive sensors, distance and height sensors, and optical methods. The review was conducted primarily by sensor technology grouped by sport, and is an imperfect approach, as at times a sport centric approach might seem more appropriate. The forthcoming sections are thus grouped by investigations into force, inertial and other which encompasses optical,

D.A. James and N. Petrone, *Sensors and Wearable Technologies in Sport*,
SpringerBriefs in Applied Sciences and Technology,
DOI 10.1007/978-981-10-0992-1_2

Table 2.1 Sensor use by publication frequency and broad usage classification from the engineering of sport conference series 2012 [11], 2002 [12]

Topic, year	Percent of total (%)	Motion capture	Video	Inertial (Acc/Gyro)	Force, strain, pressure	Wind tunnel applications
2002 (N = 113)	42	9	13	13	11	2
2012 (N = 154)	71	6	29	30	22	23

displacement and techniques that resist easy classification. An introduction to many of the sensors used here, though chiefly in a physiological context was undertaken by Cutmore et al. [13] in 2007.

Chapter 3
Sensors

Sensors in use in the field of sports will be reviewed in the following section: the quantities on which the sensors is based will be the first key of presentation, starting from load and pressure sensors and going through inertial sensors, optical sensors, angle and displacement sensors.

Temperature and humidity sensors will be considered when included into a garment. This is the new frontier of sensor technology in sports: sensors that the sport practitioner can wear or even that are embedded in the garment he/she is daily wearing and communicate with his/her portable ICT devices can compete with precise and devoted systems of sensors that are usually developed and implemented for research sessions.

A tool for orienting an early selection of most suitable sensors when addressing the measurement of typical mechanical quantities will be proposed in the following Chap. 5, dealing with Implementation.

3.1 Load and Pressure Measurement

This section looks at the measurement of loads through strain gauges, load cells and pressure measurement. The choice between a load and a pressure measurement system has to be based on several criteria that will be explained throughout the paragraph: however, it is the aim of the individual studies that primarily orients the choice. Given the fact that, in the presence of contact forces, the load in a certain direction in the integral quantity of the pressure distribution acting in that direction over the contact surface, researchers interested in the way pressure is distributed over the surface shall orient their measurement system towards pressure sensors. On the other hand, if the aim of the study is the total load acting in one or even more directions, the load measurement system shall be preferred if the local pressure distribution producing the resultant loads is not of interest.

D.A. James and N. Petrone, *Sensors and Wearable Technologies in Sport*,
SpringerBriefs in Applied Sciences and Technology,
DOI 10.1007/978-981-10-0992-1_3

3.1.1 Engineering Background

Load Measurements

Examples of Force (N), Moments (Nm) and Torque (Nm), as expressed in the more general term "Load", are widely considered in load sensors design and applications. A useful introduction on measurement of these quantities can be found in the "Instrumentation Handbook" [14], while a very detailed review of different constructive solutions for Load transducers is also recommended [15].

To use these sensors familiarity with the definitions is important, in particular pressure (normal force acting perpendicular on an element's surface, divided by its area) and shear stress (tangential force acting parallel to an element's surface, divided by its area); similarly the definition of axial strain, (as the deformation—relative change of dimension-due to the action of stress in the stress direction) and "shear strain", (as the distortion—change in the shape—of an element due to the action of shear stress.

Mechanical strain is typically measured through the deformation of resistive material in what are called strain gauges. These may be applied to particular geometries on metal to create load cells in particular directions of interest: depending on the concept of the load cell, both the axial/bending strain on particular beams or the shear strain on shear panels or diaphragms can be the mechanical quantity under measurement.

The strain on the underlying material on which the gauge is applied introduces a change in the length of the conductive grid of the gauge that is in turn producing a change in the overall electrical resistance of the gauge. These changes are usually very small and can be measured by bridges (typically the Wheatstone bridge) enabling the amplification of small resistance variations in more common voltage variations. Gauges then are connected in bridges that can have one (quarter), two (half) or four (full bridges) active branches. Increasing the number of branches increases the sensitivity of the bridge: it also brings other possible advantages like the decoupling from other undesired load components, the insulation against electrical noise, the compensation against thermal output and the possibility of compensating the influence of lead resistance.

Bridges can be applied directly to the equipment in the case of highly compliant devices (e.g. skis, paddles, tennis rackets, bicycle components) or they can be can be engineered together with a custom designed multiaxial dynamometer that can be conceived and customized for the specific application. Advantages of the latter case, such as high decoupling between channels, high number of measured components, interchangeability among different samples compete with the advantages of the former case (lower mass and cost, lower disturbing effect on the athlete). Off-the-shelf solutions (uniaxial load cells, button-size load cells, multi-axis load cells) have good utility for research, though can yield bulky or expensive component cost solutions.

In addition, we can distinguish between single or multi-component load cells (depending on the number of measured load components) and between "direct" or "indirect" load cells. If each output channel directly correlates with a desired Load component, then we have a direct load cell; if the desired load components hve to be calculated from the linear combination of the output strain gauge channels, then we have an indirect load cell. In this case, an algebraic transformation matrix has to be introduced for the recombination of measured strain channels, to build the virtual channels that will correlate directly with the desired load components by means of the calibration matrix.

An interesting example of fundamental application of strain gauge bridges in a direct biaxial load cell can be found in the octagonal ring biaxial load cell [16]. In this example (see Fig. 3.1) due to the curved beam behaviour of the two vertical sides of the octagon, a vertical load (F_Z) induces a tensile bending strain in the outer gauges 1 and 3, and a compressive bending strain in the internal gauges 2 and 4. Due to the sequence of connection in the bridge, the effect of the four gauges is amplified as Eq. (3.1) and gives the output voltage v_{out} [mV] (Ch #1) proportional to the Vertical load [N] by the direct sensitivity coefficient s_{11} of Eq. (3.2). In the case of a biaxial load cell, the second (Ch #2) channel may present a transverse sensitivity (s_{21}), when load F_z is applied.

When a shear load (F_X) is applied to the upper horizontal octagon side, due to the antimetric deformation of the octagon, tensile bending strains appears in gauges 6 and 8, and compressive bending strains in gauges 5 and 7. In this case, the output voltage V_{out} [mV] of the second channel is proportional to the horizontal load [N] by the direct sensitivity coefficient s_{22} of Eq. (3.2), and s_{12} represents the transverse sensitivity of channel #1 when load F_x is applied.

The result of the calibration process on this example of direct load cell is then expressed by Eq. (3.2), or its symbolic version Eq. (3.3): the [S] matrix is the 2×2 sensitivity matrix of the load cell. After the calibration process, usually performed statically or dynamically in a laboratory setup, when the load cell has to be used in the field for estimating the loads from the channel measurements, the calibration matrix of Eq. (3.4) has to be used. This octagonal load cell was one of the first solutions to be applied in sports engineering to measure pedal loads in cycling [32] and ground reaction forces in gait analysis labs [16]. An evident limitation is the

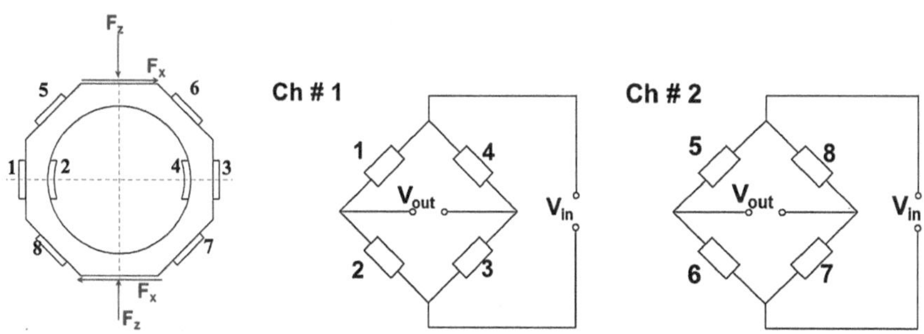

Fig. 3.1 Example of a two-channels octagonal ring dynamometer, with indication of the point of application and the bridge configuration of the two channels

need of applying the gauges to the inner part of the ring, thus preventing an extreme miniaturization of the load cell thickness. Due to such limitations, scientists also considered alternative solutions to the octagonal rings, such as the shear panel elements, in the development of multi-component dynamometers for measuring binding loads in skiing [42].

$$V_{out} \approx V_{in} \cdot \frac{K}{4} \cdot (\varepsilon_1 - \varepsilon_2 + \varepsilon_3 - \varepsilon_4) \tag{3.1}$$

$$\begin{Bmatrix} ch1 \\ ch2 \end{Bmatrix}_{[mV]} = \begin{bmatrix} s_{11} & s_{12} \\ s_{21} & s_{22} \end{bmatrix}_{[\frac{mV}{N}]} \begin{Bmatrix} Fz \\ Fx \end{Bmatrix}_{[N]} \tag{3.2}$$

$$\{ch\}_{[mV]} = [S]_{[\frac{mV}{N}]} \{L\}_{[N]} \tag{3.3}$$

$$\begin{Bmatrix} Fz \\ Fx \end{Bmatrix}_{[N]} = [S]^{-1} \begin{Bmatrix} ch1 \\ ch2 \end{Bmatrix}_{[mV]} = [C]_{[\frac{N}{mV}]} \begin{Bmatrix} ch1 \\ ch2 \end{Bmatrix}_{[mV]} \tag{3.4}$$

Force plates are examples of multicomponent load cells, they are usually constructed using a combination of four triaxial load cells applied at the four corners of the platform. The commercial ones adopted in the gait analysis laboratories are usually bulky and stiff. They are widely and routinely used in sports sciences when the action can be performed over the platform. In many cases, their dimensions and mass prevented from their direct use in the sports discipline (e.g. skiing, cycling, snowboarding) and most of the effort of researchers and engineers was devoted to the development of customized sensors having such measuring ability with compactness and lightness. Description of force platforms is beyond the scope of this review, though the principle applied as custom instrumentation is of great interest to the sports engineering community.

Other technologies in the measuring of load components include the use of piezo-capacitive materials, which creates a charge based on physical deformation, and the use of pressure sensors.

Piezo-capacitive load cells have the advantage of presenting high stiffness and high frequency response, so that their best field of application is in the field of impulsive load measurements, where impacts and shocks or vibrations occur with a high frequency content. Their main disadvantage, together with the need of a special charge amplifier, is in the possible presence of a drift during long term measurements. Care shall be taken when stability of posture over time is taken with such technologies.

Despite that, successful experiences were carried out in the field by using piezoelectric multicomponent load cells interposed between the skis and the bindings: main limitations of these experiences were the additional mass of about 5 kg per skis and the increased distance between the ski and the boot sole [17].

Pressure Measurements

Pressure sensors are a technological solution available to a researcher interested in studying how forces are being transmitted between bodies in contact. This may happen where the investigation regards the body-equipment interfaces, such as a foot-shoe interface, a buttock-saddle interface, a head-helmet interface, or when the contact between bodies is the object of the study, like a ball-racquet interface.

Pressure sensors can also be adopted as devices interposed between surfaces under contact able to integrate the pressure distribution and to give the load normal to the surfaces under contact. In this case, the limitations of the approach shall be soon highlighted, such as the spatial resolution of such pressure sensors, the variable orientation of contact surfaces, the low sampling frequency, the absence of shear stress components.

In an excellent review article, Razak et al. [18], the authors addressed the description of pressure sensors with special attention to foot plantar pressure; the paper, with further correction by Chockalingam et al. [19], gives a wide description of the technologies in use in pressure measurements, together with the quality and implementation requirements valid for such applications.

The pressure sensors can be described and classified according to their architecture, their measuring principle and their performance and implementation characteristics.

First distinction of pressure sensors can be based on their distribution of flat and rigid surfaces, to build pressure platforms, or on flexible and thin supports, to be conformed as foot soles, saddles and seats, mats and pressure bands. In the first case, the main field of application is in the field of clinical or functional analysis of gait and posture, typically in barefoot conditions, where a sound experience is associated with the study of static and dynamic footprint to evaluate for instance the pronated/supinated foot and support the design and prescription of foot orthoses. In the second case, the flexible matrix of pressure sensors can be arranged in the most variable shapes: most common are the pressure insoles, that fit into the shoes (of any type, ranging from slippers to ski-boots), the saddle and seat mats, to be used in bicycle, vehicle, wheelchairs or furniture studies, large mats to be used in bedding optimization or ulcer prevention. Further to this, it has to be clear when a flexible pressure mat contains inside a matrix of sensors, that means that the sensors tend to cover all the contact surface, or when a number of localised sensors are applied in points of the surfaces that have been proved to be significantly representative. When arranged in a flat platform, sensors usually present a higher resolution (smaller dimensions) and longer durability that flexible arrangements. On the other hand, the real pressure distribution between foot and shoe can be measured only when the foot is dynamically walking inside its shoe sole, provided that the pressure insole is thin and flexible enough to not influence the pressure distribution.

Pressure sensors are available in the market based on different measuring principles. A first group is based on capacitive a principle, where the sensor is made of two conductive electrically charged plates separated by a dielectric elastic layer: this dielectric elastic layer bends when a pressure is applied, then shortening the distance between the two plates and resulting in a voltage change proportional to the applied pressure. A second group uses resistance, when pressure is applied, the sensor measures the resistance of conductive polymer between two electrodes that decreases as the conductive layer changes under pressure. In the case of a single couple of electrodes, the force-fensing resistor (FSR) can be developed. Thirdly piezoelectric load cells, also piezoelectric pressure sensors of small diameter can be developed: in this case the sensor produces an electric field (voltage) in response to pressure, presenting some limitations again regarding the presence of drift and sensitivity to noise. Piezoresistive Sensors are made of semiconductor material: the bulk resistivity is influenced by the force or pressure applied, when the sensor is unloaded resistivity is high and when force is applied resistance decreases. Finally, most recent development of pressure sensors involves Micro Electro Mechanical Systems (MEMS), that, being embedded in the circuitry, provide a direct connection to the transmission or logging devices. This low physical dimension, low cost sensor construction method can present high-pressure range measurement capability, excellent linearity both at low and high pressures and negligible hysteresis. IN considering sporting applications solutions based on knitted fabric coated with carbon-black-filled silicon used as the strain-sensing element have been developed and applied for localized placement under the foot [26]. The sensors principle involves the strain action of wavy shaped surfaces that are pressed on the conductive fibre by the external pressure.

Comparison among sensors and technologies shall be based on the functional requirements of such sensors, that can be shortly summarized as:

Hysteresis: Hysteresis appears when the applied pressure is increased by loading or decreased by unloading and two different responses are observed: the smaller the hysteresis, the better the sensor performance.

Linearity: Sensors with low temperature sensitivity in the 20–37 °C range are preferred: changes may be due to the materials that are part of the sensor body (i.e. resistivity with temperature) as they respond differently to temperature change. Humidity can also cause changes in the sensor output and require recalibration.

Temperature and Humidity Sensitivity: The linear dependency of sensor output from the pressure applied simplifies the data post processing and is preferred.

Pressure Range: This is the key specification for a pressure sensor, to prevent "saturation" at the functional or maximal actions to be measured. Despite some maximum values are known for the applications [18], the specific values of different sports applications shall be estimated in advance, including impact or shocks.

Sensing Area of the Sensor: Size and placement of the sensor are critical for the successful measurement of peak pressures or gradients. Complexity and cost of pressure matrix raise exponentially when sensor size decreases from the usual 10 × 10 mm.

Operating Frequency: Particular care shall be taken when the analysis regards transient or impulsive actions. For example fine some researchers suggest sensors must be capable of sampling at 200 Hz [20] or 500 Hz [84].

Creep and Repeatability: As creep is the deformation of material under elevated temperature and static stress, low creep sensors are the preferred ones in pressure measurement. Repeatability, as the ability to produce reliable result even after long period of time, is highly appreciated when comparing results in a set of repeated measurements.

In order to support the reader in the comparison of different systems and solutions for in-shoe pressure measurements, seven different solutions commercially available or from recent literature presentation have been collected and compared in Table 3.1, after adapting the analogous table that can was presented by Razak [18].

The successful implementation of a foot pressure measurement system is conditioned to the fulfilment of usual requirements of a field data acquisition system, as will be highlighted later: portability (low mass), wearability (in-shoe application), wireless data transmission (no cable), low energy consumption, low cost.

3.1.2 Sport Applications

The measurement of forces, whether for sporting equipment or for human locomotion [28], has been widely undertaken in sports related research and more recently in game based fitness products [29]. Whilst traditionally used in mechanical engineering, strain gauges and load cells have been applied widely in equipment and material testing. More recently though, they have been applied to sporting equipment for locomotion studies. Sports that use equipment for locomotion have tended to be popular, presumably because of the easy of mounting equipment but also because the focus often was also on the optimization of the equipment itself.

In cycling instrumented chain ring and or cranks are replaced or retrofitted with strain gauges for bicycles that measure crank force. Jones and Passfiel [30] describe the use of a number of these technologies together with methods to calibrate them. Sanderson [31] takes a different approach, looking at the various influences on force production instead and Hull & Davis specifically at the pedal loading [32]. Force measurement on bicycles has also been used structurally to assess frame strength [33]. Strain gauges have been widely used in detecting mechanical stresses in equipment likes bicycles through commercial load cells [34] and applied forces. With the availability of low costs sensors and support instrumentation these technologies are now possible to use in the field [35]. Anecdotally there was initially a resistance to the use of these technologies in cycling because of the weight penalty (a few hundred grams), however the advantages of the information soon became clear and today they are widely adopted. Almost all professional cyclists use strain gauged power meters inbuilt in the crank that wirelessly transmit their instantaneous cycling power to a display clamped on the handlebar.

Table 3.1 Comparison of different systems and solutions for in-shoe pressure measurements (adapted from [18])

Company/ developer	Vista medical	Tekscan	Novel	Moticon	Paromed	Hong Kong polytechnic university	University Malaysia Perlis
Products	FSA® in-shoe	F-scan®	Pedar®	OpenGo Science®	ParoTec®	*Textile sensor*	*MEMS*
Reference	[21]	[22]	[23]	[24]	[25]	[26]	[27]
Technology	Resistive	Resistive	Capacitive	Capacitive	Piezoresistive	Fibers	MEMS
Thickness (mm)	2	0.15	1.9	2÷3	3.5		2
Number of sensor	128 (matrix)	960 (matrix)	99 (matrix)	13 (localized)	36 (localized)	6 (localized)	15 (localized)
Range (kPa)	260	862	1200	400	625	800	3000
Frequency (Hz)	40	750-Cable 100-Wifi	100	50	300	100	200
Hysteresis	Not Specified	24 %	<7 %	Not Specified	0.05 % at 200 kPa	<8 % at 200 kPa	Negligible
Data transfer	Cable to unit	Cable to unit	Cable and wireless	ANT wireless	Wireless	Not specified	Wireless

Rowing and to a lesser extent kayaking have widely adopted the use of strain gauges to measure force production: Secher [36] was one of the early researchers to use these sensors to examine, isometric, leg and arm strength, together with postural changes, which were compared and contrasted to other methods. Extending beyond this, Kleshnev [37] combines force and displacement sensors for a more complete biomechanical analysis. In determining placement of such sensors there are a number of accepted methods to assess force production each with limitations and advantages. Foot stretcher force and even the seat position [38], as the quasi static reaction force is a convenient and stable position to measure force production with load cells. Oarlocks are another place at the (arguable) fulcrum or load of a first class lever and more recently the force in the oar itself has been measured using an optical technique. In this technique a graded optical fibre in a distributed mode receives reflections from bend and pressure points on the fibre to build up a complete strain profile of the oar [39] using a well established technique is [40].

Force plates are widely and routinely used in sports sciences and beyond the scope of this review, though the principle applied as custom instrumentation is of great interest to the sports engineering community.

Following the pioneering work of Hull [41] and Quin et al. [42], the measurement of in-field load during skiing is still a great challenge to researchers both oriented to safety or performance improvement in skiing for measuring downhill skiers [43–45] and biomechanically similar sports such as roller skiing [46] show its extension and utility. Recently Scott (Fig. 3.2) [47] studies the skier-snow force interaction using a customised dynamometer at the binding, pressure insoles inside the ski-boots and snow pressure sensors embedded in the ski base, together with a backpack mounted Inertial Measurement Unit (IMU). Customised dynamometers that can be interposed between the boot and any type of ski binding have also been applied [48, 49] to measure the six load components developed at the bindings. Load sensors have also been used in cross country skiing to look at over-snow running surface forces [50] as well as separate heel and toe forces for athlete performance analysis [51]. Force measurement on ski poles has also been used for materials investigation and skiing technique characterization [52, 53] and investigation into the design of ice skates and performance assessment [54].

In the laboratory, measurement of force is widely used to assess the material properties of sporting equipment, usually as an aid to design. Examples include, but are not limited to, baseball bats [55], hockey sticks [56] and footwear stiffness [57]. Sporting surfaces have also been investigated [58] including investigations of various soil structures [59], as an aid to injury studies and shoe component design.

The measurement of forces has also emerged as a useful aid in the study of aerodynamics and thus useful in wind tunnels as direct measurement of drag on balls [60] and bicycles [61], though inertial sensors have also found applications here as well [62]. Fuss et al. measured the force production of climbers on holds through the use of standard forces plates [5] leading to the development of an instrumented climbing hold [63].

The indirect measurement of force, through the use pressure sensors has also been applied. The use of pressure sensors however often limits the measures to an

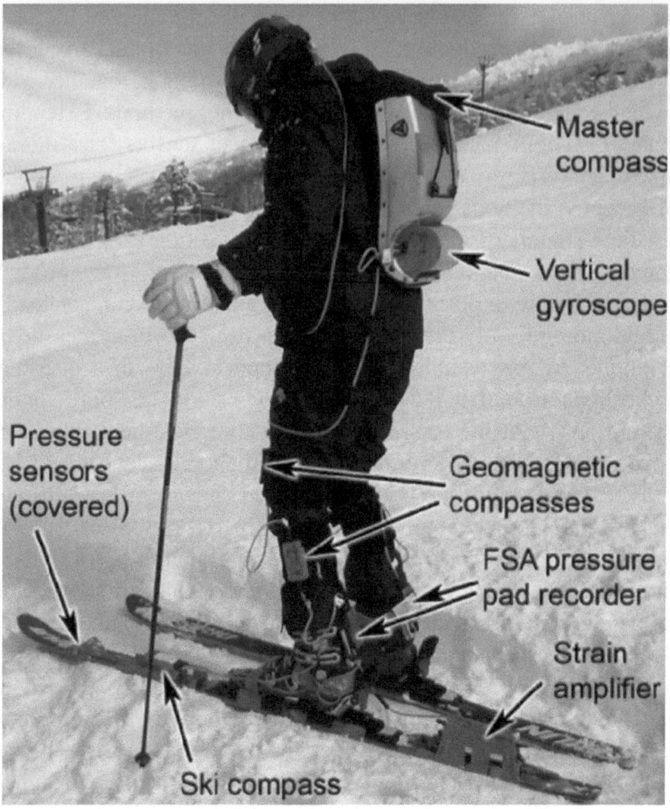

Fig. 3.2 Instrumented skier showing a variety of sensors and the challenges and complexity of integration. Used with permission [47]

aggregate quantity (rather than individual components such as shear). By measuring pressure in bands of tubing researchers have been able to evaluate the comfort of garments [64], to assess the protective elements of head equipment [65, 66], as a taekwondo scoring system [67] and for measuring the force and location of strikes in boxing [68] and related disciplines.

3.2 Inertial Sensors

Inertial sensors, commonly comprising of accelerometers and angular rate gyroscopes, measure changes in the linear and angular acceleration. More recently the inclusion of magnetometers, whilst not strictly inertial have been included as they provide a fixed reference frame. These sensors today are widely applied to the kinematics of a body were they can be used as biomechanical markers of the body activity or to derive linear or angular velocities and thus displacement and angles.

By knowing the inertial properties of the moving body, forces and moments acting on the body centre of mass and distal segments can be estimated, provided that the inertial properties are well estimated in the case of anthropometric segments [69].

3.2.1 Engineering Background

The comparatively recent miniaturisation into MEMS technology allowed to reach the size of millimetres or smaller. This process has led to their popularity as a sensor for measuring sports persons and equipment alike. Acceleration is the time derivative of velocity and velocity is the time derivative of position. Rate gyroscopes measure the angular velocity as the time derivative of angles. Thus a somewhat common and often erroneous approach is to try to obtain direct kinematic measurements, that is to numerically integrate to determine velocity and/or position. In theory provided that the initial or final position are known and that the sensors does not drift (that is an output different from zero when the acceleration or speed is physically zero) these more easily understood measures can be obtained. In practice though this has proved to be a difficult and complex task. Some of this is due to the nature of the requirements for sporting applications, namely that they are small enough to be worn together with the compromises that these make (lower accuracy and sample rates) than those used in inertial navigation systems. In these navigation applications a combination of suspended mass accelerometers and mechanical angular rate gyros (gyroscopes, including a flywheel and a motor), constitute an Inertial Navigation System (INS) where depending on their intrinsic drift, that can range from the 1°/year in the case of spacecraft applications to 1°/min or more for MEMS systems. The bulky vertical gyroscope shown in Fig. 3.2, classified as inertial grade device, may have not easily replaced by a MEMS vibrating gyro in the skiers pocket if the time of recording would have not been short enough to neglect its possible drift.

Additional to the amount of drift inbuilt in the sensors, two other requirements shall be checked when choosing a rate gyro: the maximum angular speed in °/sec and the maximum sampling frequency. When addressing the sensors on a golf club, for instance, peak values of 3000°/sec can be expected. Compromises between sample rate, resolution are often practical considerations which limit their accuracy in sporting applications and shall be dealt with in forthcoming sections through individual principles as well as design and implementation.

The widespread adoption of the mems accelerometers technology in the mass market first by the automotive industry (in the deployment of airbags) and more recently as a portable computing and game controller product [70] has led to increased on board signal conditioning and reduced capital cost and power consumption.

More recently inertial sensors oriented in the three directions have been combined with magnetometers to create a single Inertial Measurement Unit (IMU): in these applications the magnetometer senses the orientation of the unit with respect to the local magnetic field. This information, together with a triaxial accelerometer and a triaxial MEMS gyro, is used to estimate the orientation of the object: their signal enters into an iterative calculation process, historically named Kalman's filter, devoted to the compensation of the sensors drift error, enabling the calculation of the unit orientation in space [71] and the use of quaternian methods [92].

3.2.2 Sport Applications

A generalised introduction to the use of inertial sensors [72] describes the use, challenges and implementation of these sensors for athlete monitoring. Davey uses the gravitational component as a recurring reference point [73]. In addition to the gravity component Ohta et al. identifies the multiple components as translational, rotational, centrifugal and tangential [74] (summary in English [75]) and that the decoupling can be difficult.

In the dynamic sports environment, complex physical parameters are measured and observed in relation to running and stride characteristics [76] and in the determination of gait [77]. Ozaki et al. has developed a model of lower limb movement using a double pendulum model to restrict the degrees of freedom which was first applied to sport in 2012 [78]. It is thought this technique may be applied to other limb segments as well. Researchers have also used accelerometers for determining physical activity and effort undertaken by subjects. These kinematic systems have been able to offer comparable results to expensive optical based systems [79] and are particularly advantages where an ambulatory environment is important.

One of the first reported use of inertial measurements was by Wong et al. [80]. Probably the mainstay of inertial sensors has been in the area of gait analysis where researcher like Mayagoitia [79] have made significant advances in the understanding of gait through accelerometers and gyroscopes [81]. Following on from this work the applications have extended from the monitoring gait into the monitoring of whole bodies segments [82]. Reliability and validation trials include correlation to the energy systems of the body [83] through examination of runner's oxygen consumption at a variety of speeds, validation of force prediction with the use of in shoe pressure sensors [84]. GPS [9] was used to validate an inertial sensor based model to accurately predict running speeds from sensors alone.

Swimming is another area that has seen widespread application of inertial sensors. This is perhaps because of the close alignment of the capabilities of the sensors with that of the needs of the sport. Swimming is a low impact sport and thus the bodies' motion is well within the range of the capabilities of the sensors dynamic range. Swimming suffers from practical difficulties that make the analysis of swimmers motion difficult and labour intensive. These difficulties include the challenge of measuring the swimmer in a quasi static position. The use of swimming flumes has been adopted, though researchers feel the motion is not truly representative, thus video methods require both ambulatory filming and manual analysis and correlation of body movement. The application of sensors to swimming requires waterproof devices and that the drag of such does not affect the motion of the swimmer. Despite these requirements there are a number of research groups internationally working in this area. Ohgi [85] is widely acknowledged as the first to work in this area. In this paper Ohgi describes a wrist-based sensor that is able to recognise arm strokes and swimming phases. Since that time an Australian group [86] developed a device to be applied on the sacrum (on the lower back) that could recognise swim strokes and lap times to within 0.1 s of touch pad sensors and better than hand timed methods. LeSage et al. [87] developed an automated sensor system that could record and transmit sensor and derivative such information. Extension by James demonstrated a multi-segment system with video integration [88] which was applied to turn and stroke phase analysis. Swimmer feedback has also been considered in a more aggregate form [89] and an optical feedback method to the athlete considered [90]. Recently the determination of velocity of a swimmer by integration [91] and quaternion methods [92] has led to instantaneous velocity measures being able to be captured.

More detailed biomechanics has also been investigated through the use of inertial sensors. Ahmadi [93] for example used several sensors to measure individual limb segment biomechanics in tennis that corresponded with other methods such as videography [94] and proposed a tool for skill acquisition [95]. Inertial sensors have been used on sporting implements themselves including cricket [96] and as an assessment tool for illegal arm action [97]. The sensors have been used for classification and differentiation of skill in golf [98] and other swing based sports [99]. Inertial sensors have been used in instrumented boats for rowing to determine stroke events [3, 100] and in combination with GPS [101].

With advancements in technology reducing the size and weight of accelerometers, recent research has instrumented a variety of sports balls for the purposes of studying flight, release phases and impact events. Inertial sensors have been built into bowling balls [102, 103], baseballs [104, 105], cricket balls [106] as well as flying discs [107]. In each of these applications, information on spin and axis of rotation has been of interest, together with flight time and impact. Further quantification of workload in ball-based contact sports has seen the sensors used in the detection of collisions [108]. The protective element of monitoring collisions has also seen the sensors variously applied to helmets [109, 110] with a comparative analysis by Higgins et al. [111].

Jump height has been routinely measured using several inertial sensor products and found to be reliable [112–114]. In winter sports, Harding [115] applied the sensors to measure key performance metrics of jump height and rotation in snowboarding half pipe and found them to correlate well to score in international competition. An extension of his work was to use the sensors as a judging tool in international competition, where the use of the tool was also investigated for its effects on sport [116]. A pattern-based analysis of jump type has also been applied to the sport of trampolining for the detection and classification of technique [117]. Ohgi [62] used the sensors first to measure skiers motion and then the aerodynamic drag on an athlete [118]. Scott [47] used inertial sensors together with other sensors to monitor skier motion and ski forces, in a backpack solution. Brodie [8] demonstrated fusion of several technologies to provide complex measures in a simpler form to the athletes for performance assessment. The approach was successively adopted by Supej [119] using a commercial IMU suite system: Gilgien [120], on the other side, compared the estimation of skiing external forces based on a differential global navigation satellite system (GNSS) with a 3D video based approach. This paper found good relative though not absolute agreement between the methods, noting that the wearable system had significant advantages because it was easier to use in field based studies.

3.3 Optical and Other Sensors

The measurement of light directly or indirectly offers a low contact method for measurement of the human body's movement and biology. Whilst routinely used as a method of heart rate detection through photoplethysmography [13] its found application more widely in sports through light gates, sensors and specialist photography.

3.3.1 Engineering Background

Light, as a travelling electromagnetic wave has many properties that can be manipulated to convey information and can be measured easily using transducers. Transducers may measure frequency and or amplitude that can change or be made to change as a result of transmission, reflection and adsorption. Light sensors can be photo resistive, act as a diode or generate an electric potential [4], they may also be used as a direct to user feedback method.

3.3.2 Sport Applications

Light gates are a popular way to measure and control athlete speed [121], a modified version of which is to look at the stance time of runners [122]. Force production can be measured by strains placed on optical fibres changing their propagation characteristics [39]. Recently, the use of light as a transmission technique has been used to provide feedback of stroke information to a swimmer visually [90]. A variation on optical techniques is to use time of flight, such as the Doppler effect for velocity [123] or radio waves to measure speed, such as over ice velocity in a bobsled [124]. Recently, modified video that includes depth information has been applied to sports biomechanics [7], these methods have been found to have comparable reliability to conventional 3D camera analysis systems [125]. The technique uses time of flight calculations of the video array [126] and has been made popular in the Microsoft Kinect system.

3.4 Angle and Displacement Sensors

Angles between limbs or relative/absolute displacements of moving segments are typically associated to the correct execution of a sport gesture: the degree of skill and coordination, the comparison among expert, intermediate, beginner athletes is usually based on the visual capture and analysis of angles and displacements. This is the reason why this type of sensors finds wide diffusion in the field of sports assessment.

3.4.1 Engineering Background

Displacement sensors can be based on different technologies, either with contact or without. One of the most common (and low cost) displacement sensors in the field of sports applications are the "Draw-wire Sensors": these sensors measure linear movements using a highly flexible steel or polymer cable. The cable extremity is usually connected to the moving part, namely the measurement object, the housing is fixed to a stationary frame. The cable is wound around a drum, the axle of which is coupled to a potentiometer or encoder, giving respectively an Analog or Digital output signal. With a change of distance of the measurement object to the Sensor the drum rotates. This rotational movement is converted to an electrical signal and output by an encoder or potentiometer. These sensors present advantages like quasi-infinite resolution, easy, fast and flexible mounting and high reliability and service life. On the other hand, the presence of the wire can obstruct or interfere with the movement of the moving part, and the presence of a recoil spring can be inconvenient in the case of high speed of the movements. Moreover, if the displacement has more than one direction, the measurement of the single draw-wire sensor is unable to correctly locate the spatial position of the moving part, but only its radial distance along the wire.

Laser based displacement sensors present the advantages of non contact measurements systems, preserving high resolution, excellent linearity and adding high measuring frequencies. The limitations about the 3D localization of objects remain valid.

Angles expressing the absolute orientation of body segments (i.e. with respect to a fixed axis) or relative angles between segments at a body joint (i.e. angle between the reference axis embedded in the segments) are even more frequently adopted to describe the posture and angular kinematics of a subject performing an indoor or outdoor sports activity. Given their analogy with human eye perception and their prevalent role in the analysis of the execution of a complex gesture, these quantities are usually the engineering measurable quantities closest to the "trainer's eye" in the sportsground. Technique evaluation, correction and demonstration are often based (at least at an initial stage) on the evaluation and training of an absolute segment angle (e.g. trunk angle during a 100 m sprint) or of the max/min/range values of a relative angle (e.g. the relative angle at the knee during a squat-jump).

Angle sensors in a wearable sensor-setup can be then divided into absolute angle sensors (inclinometers, magnetometers, IMUs) and relative angle sensors (rotational potentiometers, rotational encoders, optic fibers sensors, strain gauge cabled sensors).

Given the improvements in accuracy and speed of Motion Capture stereo photogrammetric IR systems, they have become the validation tool for other wearable angle sensors.

3.4.2 *Sport Applications*

Displacement sensors are routinely used to detect displacement and velocity in highly constrained 1D movement in rowing applications for seat, oar and trunk movement, usually through cable based multi-turn resistive material [37] as a goniometer for joint angles [127] and as an aid in the force production of lifting weights in a gymnasium context [128]. Witters et al. [129] describes the construction of a wire velocimeter. In this technique a rotary encoder measures the paying out of a fixed line to measure distance and velocity of an athlete.

Even though these quantities have been under evaluation since the beginning of sports motion capture, (neon tubes embedded to the side of body limbs, [130] there has been always a cross fertilization (or a kind of competition) among video motion capture, stereophotogrammetry and wearable angle sensors.

Potentiometers and encoders are rotational sensors have found widespread us in sports and biomechanics studies. In a relatively recent example, Nordquist and co-authors [131] apply these sensors to ankle joints in snowboarding (Fig. 3.3), other techniques include successive substitution in some 2D applications by

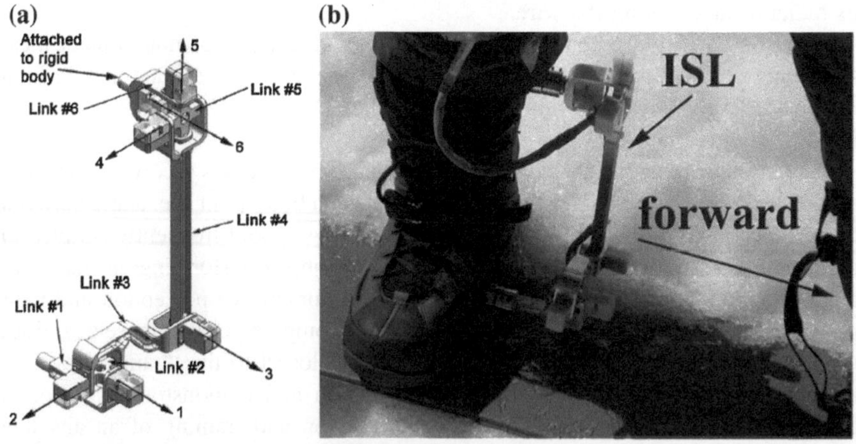

Fig. 3.3 The instrumented spatial linkage (ISL) based on 6 links and 6 rotational potentiometers: the design representation (**a**) and the field application on a snowboard test (**b**) [131]

biplanar angle sensors [132]. A likely trend is that relative angle sensors will be progressively substituted by the combination of two absolute sets of angles calculated by means of two IMU's [97, 133], accuracy is highly dependant on sensors used and the range and speed of the motion desired to be measured. The advantage is the absence of cumbersome and un-rugged spatial linkages, the absence of cables or wires crossing the joint, the possibility of including the IMU's (of smaller and smaller dimensions) in a wearable suite.

3.5 Garment and Apparel

Whilst garment, design and fabrics might seem well beyond the scope of this article, the very recent inclusion of sensors into garments, when coupled with the progressive miniaturisation of sensors and electronics makes it at least worthy of inclusion. In the last few decades, the development of sport garments and apparel has come to include the use of sensors. A wider extension of this is the concept of wearable technology which will be discussed later. Sensors are either used for inclusion in the garment itself or as an aid to garment design. For example measurements for aerodynamic or hydrodynamic assessment [134], as well as of the perceived and measurable degree of hygro-thermal or stretch-pressure comfort associated to a garment.

3.5.1 Sport Applications

Examples of possible applications range from localized wired temperature sensors (thermocouples) in a laboratory setting to ingestible telemetric core body temperature sensors [135, 136] to be used in the field.

Hofer [137] and coworkers used a miniaturized wired temperature and humidity sensor to explore the behavior of ski boots in relation to temperature and humidity, at 3 locations on the foot. Colonna et al. [138] used a wireless sensor [17 mm diameter, 6 mm thickness] embedded in the liner sole of the skiboot to compare the effect of liner during real skiing or simulated skiing sessions in a climatic chamber: in this case, the advantage of wireless connection competed with the disadvantage of larger sensor dimensions.

In-ear thermometer enabling to measure the core body temperature at the acoustic meatus was used to compare two jackets with different liners composition [139] during climatic chamber simulated cycling tests: on the contrary, core body temperature was measured in the field during a sailing session by means of an ingestible pill sensors [136] with the aim of validating a model of clothing thermal conduction given the external ambient conditions. Ingestible sensors are much less invasive than unpleasant rectal temperature sensors [140]. The validation of predictive thermoregulation models seems the main justification to the use of diffused hygro-thermal sensors or of core body temperature sensors [141], the parallel correlation with the degree of hygro-thermal comfort perceived by sport endeavors is a further field of progressing research.

The attention to textiles and garment stretch properties in the field of sport suites, leading to possible enhancements of muscular efficiency, drive the research towards strain sensors [142] that can be embedded in the fabric while still transmitting signals correlated to the stretch applied to the garment.

Embedding of sensors typically uses embedded wires and more recently printed conductors have also been explored [143]. Novel materials have also been investigated these include the use of graphene as strain [144] and torsion sensors [145], its application to flexible electronics [146]. The body worn environment is an emerging area of textile chemistry and textile technology is developing a multitude of textile based sensors, applicable to sports and health science.

Chapter 4
Approaches

Approaches to the application of sensors in sports tend to follow a number of paths. Engineers and technologists tend to start with a technology and apply it to a sport; ideally there is a perceived problem to solve or an interesting opportunity to gather data. Sport professionals (athletes, trainers, instructors.) tend to have a specific problem they are trying to address and begin with a known technology to develop a solution. These are very good approaches provided the engineer's perception of a problem to solve is well grounded and the sports professionals' knowledge of technology is broad. Both approaches may benefit where there is a translational opportunity to take a successful technology application from other disciplines (e.g. a physiology measure from medicine), an emerging new technology, or a technology successfully used in another sport.

While this work serves to inform about the various technologies and application opportunities in sport, the wider literature also has examples of technologies that are applied to sports where there is some misalignment: this is a value judgement and left to the reader to consider. This misalignment tends to arise where there is a mismatch between the technology and a perceived need, or where a particular technology is perceived as a solution and applied inappropriately.

While there is no recipe or generic approach to implementation, there are several strategies and methodologies that can aid the implementation of sensors in sport and the innovation process in general [147]. User centred design can help identify the needs [148, 149] although it does not necessarily identify the technology required. A multi-stakeholder needs analysis [150] is a recent approach useful in the wider content of technology (and other) innovation. Delft University have developed a resource for the entire design process from which particular methodologies of approaches may resonate [151]. Within each of the described methodologies whether it be wanting to apply a particular technology, work within a particular sport or identifying targeted needs, a discovery or agile [152] based approach can be helpful. By applying sensors, or a simulated analogue of them, to the intended application in a rudimentary form, quick development cycles can yield the desired

D.A. James and N. Petrone, *Sensors and Wearable Technologies in Sport*,
SpringerBriefs in Applied Sciences and Technology,
DOI 10.1007/978-981-10-0992-1_4

and often serendipitous information, from which a cohesive research and development plan can be developed.

Given the competitive nature of sports, where time is a strategic factor, an effective approach to the adoption of sensors in sports that has proven successful also in several other fields of engineering can be based on the statement that '*a good design tested now is better than an optimal design tested a year later!*'.

Chapter 5
Implementation

The application and development of sensors to the sporting, or any context for that matter is complex and is systematically addressed in this section. The sporting context brings unique challenges, for example, the harsh environment, electrically and physically, the demand of small dimensions and minimal power consumption, the increasing request of real-time feedback to use.

Sensors and accompanying instrumentation are subjected to physical shock, often encountered through physical movement and contact forces, and what could be considered a quasi-aquatic environment through exposure to the elements, including perspiration. Significant temperature variation from the sun/shade environment, body contact and wind chill are also considerations.

In this section, consideration of the sensor properties, appropriate conditioning of the sensor output, power demands and the conversion and transmission of sensor data in digital form are introduced.

Figure 5.1 shows the various building blocks and eventual flow of information to the end user, which will be discussed in.

5.1 Sensor Selection and Characteristics

Sensors ideally are the conversion of a desired physical quantity we want to measure into an electrical signal. However in practice sensors exhibit non-ideal characteristics (see Fig. 5.2) such as non-linearity, hysteresis (where the sensor behaves differently depending on if its value is increasing or decreasing), noise, offset, sensitivity to other phenomena such as temperature and degradation over the sensors lifespan. D'Amico et al. [153] provides a useful introduction to the topic. Fortunately with care, calibration and post processing acceptable results are often obtainable.

D.A. James and N. Petrone, *Sensors and Wearable Technologies in Sport*,
SpringerBriefs in Applied Sciences and Technology,
DOI 10.1007/978-981-10-0992-1_5

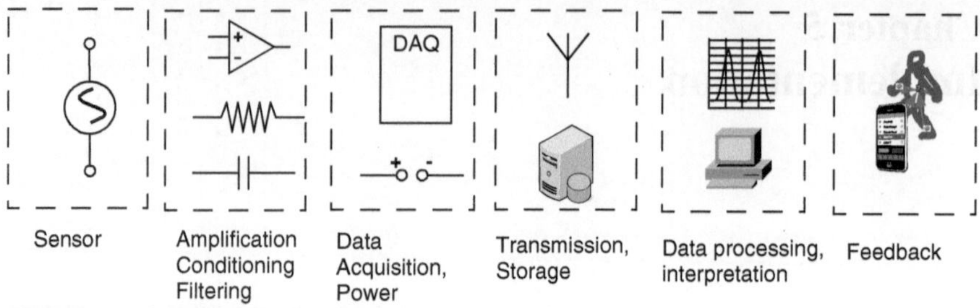

Fig. 5.1 Sensor application building blocks

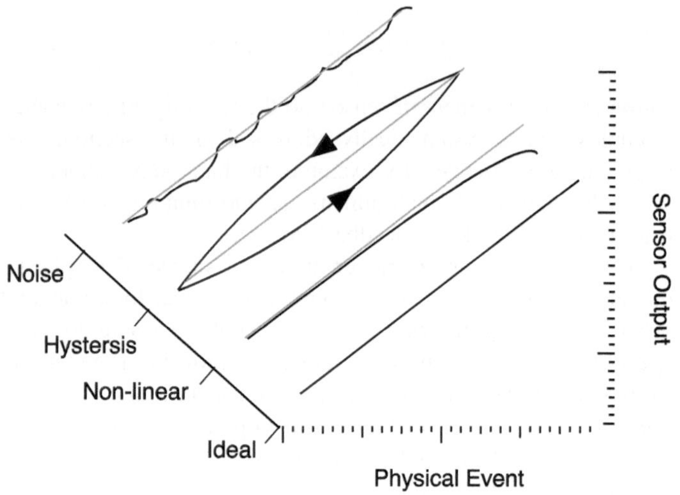

Fig. 5.2 Sensor response, idealised and non-ideal characteristics

Figure 5.2 shows some characteristics as deviations from an ideal linear sensor conversion. Sensor characteristics are usually available in the form of an accompanying data sheet, along with the manufacturers definitions, which describe the sensors requirements, responses and limitations. Additionally data sheets often contain sample application notes, reference circuit designs for sample usage and can be quite helpful as a starting point. A note of caution, all electronics products, including sensors, usually have a limited manufacturing lifespan, checking not only current, but future availability maybe an important consideration.

In addition to the characteristics of the sensors themselves, sensors measure the physical environment at their location, thus consideration of the placement environment and design is required to ensure that the location is representative of what is to be measured. For example the sensors on a bike-rider interface should measure characteristics of the bike rider interaction [35, 154] without interfering with or

Table 5.1 Indication of most suitable sensors compared to typical mechanical quantities: a tool to support early sensor selection

	Type of Mechanical Quantities										
	Kinematic							Kinetic			
	Position	Speed	Acceleration	Absolute Angle	Relative Angle	Angular Velocity	Angular Acceleration	Load	Moment	Pressure	Surface Area
SENSOR TYPE	x (y, z) [m]	v [m/s]	a [m/s²]	Angle θ [rad]	Angle φ [rad]	ω [rad/s]	α [rad/s²]	F [N]	M [Nm]	p [kPa]	A [cm²]
WIRE ENCODER or POTENTIOMETER	DIRECT	Deriv.	Double Deriv.	-		-	-	-	-	-	-
LIGHT BARRIER	-	DIRECT	-	-		-	-	-	-	-	-
LASER / DOPPLER VELOCIMETER	Integr.	DIRECT	-	-		-	-	-	-	-	-
ACCELEROMETER	Double Integr.	Integr.	DIRECT	Static		-	-	*Mass Derived*	-	-	-
INCLINOMETER	-	-	-	DIRECT	*By Difference*	Deriv.	Double Deriv.	-	-	-	-
GONIOMETER	-	-	-		DIRECT	Deriv.	Double Deriv.	-	-	-	-
GYRO	-	-	-	Integr.	*By Difference*	DIRECT	Deriv.	-	*Inertia Derived*	-	-
Inertial Motion Unit IMU	-	Integr.	DIRECT	DIRECT Magn.	*By Difference*	DIRECT	Deriv.	*Mass Derived*	*Inertia Derived*	-	-
LOAD CELL Uniaxial / Multiaxial	-	-	-	-		-	-	DIRECT	-	-	-
TORSIOMETER [Nm]	-	-	-	-		-	-	-	DIRECT	-	-
FORCE PLATFORM	-	-	-	-		-	-	DIRECT Fx, Fy, Fx	DIRECT Mx,My,Mz	-	-
Single PRESSURE sensor (Localized Force Sensor)	-	-	-	-		-	-	Integr. F norm.	-	DIRECT	-
PRESSURE Platform (Matrix of pressure sensors)	-	-	-	-		-	-	Integr. F vert.	-	DIRECT	DIRECT
PRESSURE Mat or Insole (Matrix or Localized)	-	-	-	-		-	-	Integr. F norm.	-	DIRECT	DIRECT

LEGENDA: *Deriv.* Derivation; *Integr.* Integration; *Magn.* Magnetometer; *Norm.* normal to surface; *Vert.* Vertical

disturbing the phenomena, a temperature sensor measuring atmospheric temperature should not have any contribution from the athlete it is attached to and a heart rate monitor would be free of the muscle generated electrical signal (EMG) from limb movement [155]. Conditioning of the sensor signal can be an aid in the reduction of unwanted interferences and noise and is of prime importance to ensure data integrity.

As a support to the choice of sensors in the measurement of different mechanical quantities that can be of interest in the field of sports engineering, Table 5.1 presents the rates sensors compared to desired typical mechanical measures. When the sensors gives the "Direct" measurement of the corresponding quantity, the matching can be considered as optimal and attention has to be placed to the other accuracy and performance characteristics of the sensor. When the quantity is obtained by Derivation, care has to be taken about the risk on nosy signal outcomes; when Integration is performed, the risk is of an unrealistic drift of the outcome due to incorrect or unstable zeroing of the source signal.

5.2 Signal Conditioning

Conditioning of the sensor signal involves knowledge of the sensors characteristics, the application environment as well as any external sources of noise, physical or electrical artefact. Here appropriate amplification of the signal; removal of bias voltage and correction of the signal can be undertaken. Commonly conditioning usually involves filtering whether at the sensor or later in a digital manner. Filtering is often used to remove voltage offsets, high or low frequency noise and frequently the presence of main power interference (50/60 Hz). A treatment on signal conditioning, with a focus on the sources of noise in sensors [13] is given by Cutmore et al. [156], which has a particular emphasis on small signal conditioning for physiological based sensors. The article outlines simple steps, which can be taken to remove sources of noise, together with important design decisions to minimise or reduce the presence of noise, considerable attention is placed on the use of filters. Filters, working in the frequency domain are usually defined by a particular cut-off frequency and order, which determines the frequency and how severe the filter is. Careful selection of the cut-off frequency so that it does not remove the signals of interest is important and the order of the roll-off (usually in decibels per decade) determines the severity of the signal filtering. A useful rule of thumb in a research context is to go a decade above the signal of interest, though commercial devices may approach or even be lower that the signal of interest. For example in gait studies it is widely accepted that up to 10 Hz is enough to capture information from movement sensors, though typically digital sample rates are an order of magnitude higher (from 100 Hz) to preserve higher order information [83]. Among the many popular engineering texts for circuit design are those that which favour a practical approach, and are well suited to people newer to electronics [157]. Recently the complexity of sensors has increased such that many have on board signal conditioning and noise reduction circuits (see an IMU data sheet for an example [158]): adequate power however is an important aspect to consider.

5.3 Power

While the size of sensors, support electronics and batteries has continued to shrink in recent years the power budget is an essential design element: researchers have to consider that acceptable size, operating time and data quality are inevitable trade-offs with each other. For example in a number of applications the size of the battery can be larger than that of the sensor and electronics combined. Individual components may require different supply voltages so step up or down transformers may be required. Whilst the SI unit of power is Watts, typically power consumption in sensor applications is discussed in terms of current (usually mA or μA), and battery capacity is often quoted as mAh where 100 mAh is a current draw of 100 mA for an hour, or 10 h at 10 mA.

Table 5.2 Sample power budget for a sensor system

Item function	Description	CODE	Current (inactive, active)
Microcontroller	ATMEL low-power 8-Bit	AT90USB1286	1 μA, 25 mA (typical)
Accelerometer	MEMS 3-axis	(LIS331DLH)	10 μA, 250 μA
Gyroscope	A MEMS pitch and roll gyroscope (2 channels)	ITG-3200	5 μA, 6.5 mA
LCD Display	Liquid Crystal display and driver	UG-9664HDDAG01, SSD133	25 mA
Radio transceiver	Nordic 2.4 GHz	nRF24L01+	0, 14 mA
Battery	Lithium-polymer cell	Battery	120 mAH
SD memory card	Non volatile memory	SD memory card	0, 10 mA (start-up, write)

Table 5.2 shows a typical power budget for a previously described sensor system [88].

Clearly there are a range of design options and decisions that can affect power consumption (current draw) of the sensor system. Thus careful consideration of the operating environment is required, for example if high-speed sampling is required then either a higher capacity battery is required which will affect the size of the final unit or a shorter run time is mandated. Depending on the application it is possible to have a smaller device operational for hours [88] or months [159].

5.4 Data Acquisition and Memory

Memory storage in particular capacity is less of a concern with access to cheap robust memory storage cards, however trade-offs between sample rate and data resolution are important considerations for both power and internal resources. A desirable investigation is often to sample data initially at a maximum sample rate and resolution. This is however a high cost option in processing power, current consumption and storage capacity. Often, with knowledge of the application area both the resolution and sample rate can be more finely tuned. Temporal access to memory and data buffers needs to be considered carefully as the load on the sampling system may introduce variability. Solutions include interrupt driven sampling [88], single, double or circular buffering [87] and limiting the write cycle to memory to when there is sufficient data to write. Pre-processing of data, for example basic filtering or compression, prior to storage can lower the cost of storage (write cycle and write time).

5.5 Wireless

Real time access to sensor information and feedback has great value in sports and has been greatly facilitated by the advent of low cost wireless technologies. Wireless technologies have advanced considerably in recent years giving rise to many communications protocols that target the many different balances between throughput, power consumption and complexity. There are essentially two approaches to implementing wireless solutions: utilising an off the shelf existing wireless protocol or custom designing a wireless protocol for a specific application. Existing protocols and products significantly reduce the development time of a wireless sensor network and more time can be dedicated to analysis of the retrieved data. Custom designs on the other hand can often create a more specific and hence better overall solution. Given project lifespan for many research endeavours it is likely that any chosen technology will be updated, out of date or obsolete by the time any solution is complete.

Existing wireless protocols and products can be global standards ratified to a standards body or proprietary wireless networks implemented by a single company. The risk of using proprietary radio standards to implement a wireless sensor network solution is that support and supply of radios can be limited as only one company produces them. On the other hand, members pass the ratified standards from many companies, which ensure that the specification is robust and usually ensures that many vendor options are available.

The Institute of Electrical and Electronics Engineers (IEEE) Standards Association has ratified a number of different categories of wireless network standards [160] that range in power in complexity from computer network standards, low power and bandwidth protocols (like Bluetooth) and self aligning networks. An example of intrest such as wireless body area networks is contained within the IEEE802.11.6 substandard. Manufacturers like Nordic have simpler protocols where there is flexibility in implementation such as the previously described swimming sensor [88]. Recently a Bluetooth like protocol called ANT [161], among others has found popularity in the low power applications market and been adopted by many manufacturers, particularly in the areas of sport. Whereas Bluetooth utilises a star master/slave topology, ANT utilises multiple channels thus avoiding limitations on network size. Careful consideration to antenna use, power requirements and operating time are important consideration when choosing both products and protocols as battery size is often much larger than the instrumentation and sensors combined.

Communications in wireless sensor networks typically operate in one of a number of unlicensed communications bands such as the Industrial, Scientific and Medical (ISM) bands, which vary from country to country in frequency and allowable power. Thus communications in these frequency bands must conform to strict transmit power and bandwidth guidelines that are regulated by each country's government regulatory bodies and it is necessary to research the relevant communication laws, particularly where international collaboration is considered. For example something as simple as changing the antenna on a wireless communications module can cause a wireless network operating within the legal guidelines to breach the power regulations set for that frequency band.

If the sensor system is set to transmit the information wirelessly then consideration of what technology to use is a complex question. Is the realistically available bandwidth for transmission sufficient for the intended application? This is a complex question as the noise of the environment (from other devices), the sharing of the bandwidth with multiple devices for example must be taken into consideration. How far the information must be transmitted also has an effect on transmission power, current draw and acceptable antenna size must be considered. Specifications for wireless devices often quote the maximum data rates and these are often best case and include the 'padding' of error correction bytes and network protocols and thus may be able to carry considerably less information [162].

The "Field Lab" European network [163], the outcome of a recent EU funding project adopted across several countries of North West Europe, can be mentioned as an example of an emerging approach to promote and encourage daily life sports activity. This can be seen as an example of a demand-pull process asking for the setting a wireless permanent system of data collection and storing with the possibility of free access to users will enable the sports endeavors to self-collect data using their smartphones: there are already examples of such installations in the tracks for athlete training purposes [164].

5.6 Data Processing

The use of sensors has the potential to create large data sets, which require subsequent analysis. For example the sensor system described in Table 5.2 makes use of 6 sensors at a sample rate of 100 Hz with 16 bit resolution. In this example sampling for an hour produces 4 Mbytes of data. Whilst this isn't a large data set by modern standards, if multiple subjects, multiple monitoring points, test sets with repetitions and longitudinal studies are conducted the size can quickly grow. Data reduction is a common technique employed to reduce the amount of data, in such cases pre-processing of the data on board, such as down sampling the data, or data aggregation where one sensor is used in common mode to calibrate another can reduce the effective sample rate and number of sensors. Other techniques include reporting the numerical difference between signals rather than absolute values, both as an aid for storage as well as for data transmission.

Post processing, at least in the research context, is usually undertaken using personal computing solutions, though embedded processing is becoming more popular [70] and is designed for the specific needs of the work at hand. Data processing design decisions depend somewhat on the level of custom hardware developed and whether it will be used routinely or in a research context. A select few of high-end off the shelf data acquisition hardware platforms (such as those from National Instruments, HBM, Data Translation, Kistler…), are accompanied with sample analysis tools. Beyond this the use of symbolic language such as Matlab is popular [165] and spreadsheet packages such as Excel are routinely used and a great common platform to use with sports scientists. Beyond this custom software developed in languages such as, but not limited to, Visual Basic, Java and C/C++ have increased design time, with the advantage of better code and run time efficiency.

5.7 Feedback

The possibility of giving a real-time feedback to the users is attractive to athletes, trainers, amateurs and represents a rising "demand pull" driving towards the development and implementation of sensors in sports. The way feedback can be delivered can vary as the various sensing and perception abilities of human being: from visible light communication in swimming [166], through sonification [167] in rowing, to visual and sound feedback by means of a tablet screen in golfing [168], rowing [169] and rehabilitation training [170].

The combination of pressure sensors in the ski-boot insoles, Bluetooth communication, smartphone interface and earplug real-time feedback with indications about the correct skiing technique [171] is a fascinating example of feedback applications from sensor measurement using a smartphone platform for coaching or self-training.

It is worthy to observe, however, that neither the presence of the hardware components of such a system nor the software or firmware implementation itself ensure a meaningful and successful application of the technology. The key passage of data to useful information to sports relevant information is the real knowledge process as an aid to motor learning and skill acquisition [172]. Thus, the core knowledge in sports engineering is to find the correct algorithm transforming the measurements into functional evaluations or training instruction: this comes after the development of the technology system its reliable implementation in the sport (and thus market) at accessible cost. There's still a lot to do for teams of sports engineers and sports scientists in this direction: not only reducing the time-delay for the feedback, but building knowledge from reliable data.

5.8 Packaging

Packaging refers to how the developed sensor or sensors are housed or contained. This is often the last stage of the design hardware lifecycle process and may be custom built or no more than a convenient box. Design solutions that include end user analysis from the beginning [143] with consideration for the available technologies are often more sophisticated and considered more ideal.

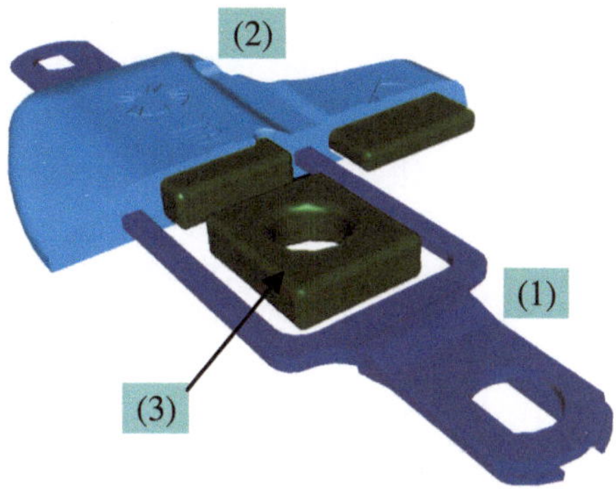

Fig. 5.3 Packaging for a wearable swimming technology [175], showing an outer flexible casing (*1*), skeleton (*2*) and potted circuit modules (*3*)

Usually electronic components and solutions are designed with an outer casing, housing electrical components sensors and circuit boards, which can be encapsulated at component or at circuit level using physical mounting or potting in resin or ceramic. The potting or encapsulation is undertaken to protect components from fatigue or failure in the harsh environment and to make the device more user friendly. Minimum clearance requirements, attachment to external circuits and sensors via connectors and housing requirements such as waterproofing are important considerations. There is considerable variety in packaging from a one off prototype in a standard instrument case, to form fitting encapsulated solutions and more recently integrated clothing solutions [173]. The costs of solutions vary as setup and production costs need to be carefully considered. For many research needs, single or small quantity solutions often necessitate the use of off the shelf solutions, though milling or 3D printing, is often attractive. The recent rise of 3D printing as a cheap alternative for one off, low volume or even small manufacturing runs is advantageous because of the lower per unit costs, ease of customisation and complex designs that can be printed [174].

As an example of a customised package for sporting application Fig. 5.3 shows a potted hip mounted swimming sensor in a flexible material, shaped by a custom built mould. A strap was added for physical attachment to the athlete via a waist belt. In this example a variety of manufacturing techniques were explored, together with their respective costs and design complexities. The final design implemented was a stereo lithography (SLA) mould. Flexible materials were trialled including latex rubber, silicone elastomers and polyurethanes. 40A Poly Urethane was found to be the most suitable for strength and flexibility.

Chapter 6
Future Directions

The continual development, invention and refinement of sensors has played a large role in their adoption for the sports environment to date and is likely to continue for the foreseeable future. This development is under pinned by a number of trends. These trends include miniaturisation, the increased metrification of sport, the rise of wearables as consumer products, market forces, technology convergence and their wider utility in related disciplines such as health.

Progressive miniaturisation of microelectronics, sensors and the increase of complexity of most technologies has followed the maximum of Moore's law. Moore's law was an observation in the mid 1960's that "the number of transistors on integrated circuits doubles approximately every 18 months" [176] and has largely held true to this day. Moore's observation has found wider utility as a generalised predictor of technology trends such as speed, complexity and market size as well. For any given complexity of a device (such as a sensor), it will become progressively smaller, requiring less power and probably becoming cheaper and lower powered. This has certainly been the case for the accelerometer and other MEMS devices [2] which are now just a few millimetres in size and consume in the order of microamps rather than hundreds of milliamps at their inception. In the fields of sports science and sports engineering where size is critical to adoption, this progressive reduction of size has seen the growing application of these technologies into the sporting realm, as well as in health care [177].

Arising from the more widespread adoption of sensors there has been a corresponding need for data analytics and interpretation, for example many professional football codes in both the training and competition environment [178]. This increased metrification of sports persons continues to grow as the user base grows in sophistication and appetite. In this application space the gathering of sensor data, processing and storage remotely offers many opportunities for the wider usage and collection of data well beyond the athlete and equipment needs including improving the spectator and media experience. Whilst variously labelled as the quantified self,

D.A. James and N. Petrone, *Sensors and Wearable Technologies in Sport*,
SpringerBriefs in Applied Sciences and Technology,
DOI 10.1007/978-981-10-0992-1_6

data analytics, big data, cloud computing and the Internet of Things (IoT) [179] this field is likely to change rapidly and one or several new disciplines will emerge from this.

The concept of wearable technology is not new though with recent advances in technology and the convergence of common consumer technologies into single small technologies it is now a term that is somewhat in vogue. It has captured the imagination of many and is seen widely as a growth opportunity for sports as consumer lifestyle products and is a growing intrest from the health sector. Further supporting the development a range of drivers of technology change such as defense, space exploration [180], elite sport [181], and medical [182, 183] from whom many needs and innovations have come.

Wearable technology implies both something that is worn and something that is technological. We consider something is worn if it is routinely worn daily or perhaps only removed at night for sleeping. While wearable traditionally might mean all items of clothing, it includes today heart rate sensors, wristwatches and personal electronics, more recently the very personal attachment of users to their smart phones [184] highlights that they need not be small. Technology too has many definitions and while we have examined sensors and electronic technologies chiefly, the intersection with materials science in particular is emerging. Materials in textiles are good candidates as a substrate for supporting electronic technologies, such as bionic skin and as sensors themselves [185]. By combining internal component housing and fortification with external packaging [186] size can be further reduced. New technologies under development though yet to mature are coming from the nanosciences [187].

At the time of writing global annual growth rates for wearables exceeding 20 % are widely reported as is increasing integration with web technologies [188], with market valuations in excess of $20B USD not uncommon [189]. Wider trends showing significant growth in the number of smart devices on the planet [188] and double-digit growth in wearables [190] are also reported and likely to have a positive effect.

In the consumer electronics space there has been both the widespread adoption of sensors into many products, along with the convergence of products that have common elements with each other. Common elements include computational processing power, sensors for user interface and display screens have combined discrete technologies such as cameras, portable computing and mobile phones into single platforms. These platforms are currently labelled smart phones, smart watches and are themselves used as sensor platforms in sports applications [191].

Chapter 7
Conclusions

The use of sensors has a vital role to play in the sporting context, for assessing and developing new equipment, as an aid to sporting performance and injury reduction. These areas are seeing considerable interest and growth. This work has explored the current and emerging trends in sensors and sought to highlight some of the developments going forward. While there are many examples of sensor technology use in sport, routine or novel, our focus has been on the usage through research developments and emerging trends, though some attention to commonplace technologies has also been included. Drawing on the literature both the approaches and principals for the use of sensors in sport have been outlined, together with references to key works, it is hoped that the reader finds this useful in embarking on such endeavours. The development of technologies is fast paced and accompanying that is an exponential growth in the use and development of computing resources, thus while the work is comprehensive on content not all works can be included and given publication times will inevitably date. However it is hoped that illumination through trends, examples and principles for adoption will be an aid for anyone considering the use of sensors in sports.

D.A. James and N. Petrone, *Sensors and Wearable Technologies in Sport*,
SpringerBriefs in Applied Sciences and Technology,
DOI 10.1007/978-981-10-0992-1_7

References

1. Panizzolo FA, Marcolin G, Petrone N (2013) Comparative evaluation of two skiing simulators as functional training devices for recreational skiers. J Sports Sci Med 12(1):151–158
2. Walter PL (1997) The history of the accelerometer. Sound vib 31(3):16–23
3. James DA, Davey N, Rice T (2004) An accelerometer based sensor platform for insitu elite athlete performance analysis. IEEE Sens, Vienna
4. Diefenderfer A, Holton B (1994) Principles of electronic instrumentation. Saunders, Philadelphia
5. Fuss FK, Niegl G (2010) Biomechanics of the two-handed dyno technique for sport climbing. Sports Eng 13(1):19–30
6. Choppin S, Goodwill S, Haake S (2011) Impact characteristics of the ball and racket during play at the Wimbledon qualifying tournament. Sports eng 13(4):163–170
7. Choppin S, Wheat J (2013) The potential of the Microsoft Kinect in sports analysis and biomechanics. Sports Technol 6(2):78–85
8. Brodie M, Walmsley A, Page W (2008) Fusion motion capture: a prototype system using inertial measurement units and GPS for the biomechanical analysis of ski racing. Sports Technol 1(1):17–28
9. Neville J, Rowlands D, Wixted A, James DA (2011) Determining over ground running speed using inertial sensors. Procedia Eng 13:487–492
10. Pope N, Kuhn KA, Forster J (2009) Digital sport for performance enhancement and competitive evolution: intelligent gaming technologies. Information Science Reference IGI Global
11. Drane P, Sherwood J (2102) The engineering of sport conference 2012, Procedia Eng 34: 1–890 Elsevier
12. Goodwill S, Haake S (2002) The engineering of sport 4. Blackwall, United Kingdom
13. Cutmore T, James DA (2007) Sensors and sensor systems for psychophysiological monitoring: a review of current trends. J Psychophysiol 21(1):51–71
14. Boyes W (2009) Instrumentation reference book. Butterworth-Heinemann
15. Stefanescu DM (2011) Handbook of force transducers: principles and components, Springer, ISBN 3642182968, 9783642182969
16. Hull ML, Brewer R, Hawkins D (1995) A new force plate design incorporating octagonal strain rings. J Appl Biomech 11:311–321
17. Federolf P, Roos M, Lüthi A, Dual J (2010) Finite element simulation of the ski–snow interaction of an alpine ski in a carved turn. Sports Eng 12(3):123–133
18. Abdul Razak AH, Zayegh A, Begg RK, Wahab Y (2012) Foot plantar pressure measurement system: a review. Sensors 12(7):9884–9912
19. Chockalingam N, Healy A, Naemi R, Burgess-Walker P, Razak AHA, Zayegh A, Wahab Y (2013) Comments and reply to: foot plantar pressure measurement system: a review. Sensors 2012 12:9884–9912. Sensors 13(3):3527–3529

D.A. James and N. Petrone, *Sensors and Wearable Technologies in Sport*,
SpringerBriefs in Applied Sciences and Technology,
DOI 10.1007/978-981-10-0992-1

20. Urry S (1999) Plantar pressure-measurement sensors. Meas Sci Technol 10. doi:10.1088/0957-0233/10/1/017
21. Vista Medical, FSA in-shoe documentation. Available online: http://www.pressuremapping.com/index.cfm?pageID=13&section=25. Accessed Aug 2015
22. Tekscan, Tactile Pressure Measurement, Pressure Mapping Systems, Force Sensors and Measurement Systems. Available online: https://www.tekscan.com/products-solutions/systems/f-scan-system?tab=configuration Accessed Aug 2015
23. Novel Quality in Measurement http://www.novel.de/novelcontent/pedar. Accessed Aug 2015
24. OpenGo Science booklet Available online at: http://www.moticon.de/support/documentation. Accessed Aug 2015
25. paromed. Products. Foot presssure meas. Parotec. http://www.paromed.biz/paroTec-3-2.html Accessed Aug 2015
26. Shu L, Hua T, Wang Y, Li Q, Feng DD, Tao X (2010) In-shoe plantar pressure measurement and analysis system based on fabric pressure sensing array. Info Technol Biomed IEEE Trans 14(3):767–775
27. Wahab Y, Zayegh A, Begg RK, Veljanovski R (2008) Design of MEMS biomedical pressure sensor for gait analysis. In: Proceeding of IEEE international conference on semiconductor electronics, 2008 (ICSE), Johor Bahru, Malaysia, 25–27 Nov 2008, pp 166–169
28. Billing DC, Nagarajah CR, Hayes JP, Baker J (2006) Predicting ground reaction forces in running using micro-sensors and neural networks. Sports Eng 9(1):15–27
29. Clark RA, Bryant AL, Pua Y, McCrory P, Bennell K, Hunt M (2010) Validity and reliability of the Nintendo Wii Balance Board for assessment of standing balance. Gait posture 31(3):307–310
30. Jones SLP, Passfield L (1998) The dynamic calibration of bicycle power measuring cranks. The engineering of sport, pp 265–274
31. Sanderson DJ, Hennig EM, Black AH (2000) The influence of cadence and power output on force application and in-shoe pressure distribution during cycling by competitive and recreational cyclists. J sports sci 18(3):173–181
32. Hull ML, Davis RR (1981) Measurement of pedal loading in bicycling: I. Instrumentation. J biomech 14:843–855
33. Manolova AV, Bertucci W (2011) Measurements of strain in bicycle frame through different conditions of test. Comput Methods Biomech Biomed Eng 14(1):57–59
34. Petrone N, Susmel L (2003) Biaxial testing and analysis of bicycle-welded components for the definition of a safety standard. Fatigue Fract Eng Mater Struct 26(6):491–505
35. Petrone N, Giubilato F, Giro A, Mutinelli N (2012) Development of instrumented downhill bicycle components for field data collection. Procedia Eng 34:514–519
36. Secher NH (1975) Isometric rowing strength of experienced and inexperienced oarsmen. Med Sci Sports 7(4):280–283
37. Kleshnev V (2002) Power in rowing. Int res sports Biomech 224–230
38. Nilsson JE, Rosdahl HG (2013) New devices for measuring forces on the Kayak foot-bar and on the seat during flat-water Kayak paddling. Int J Sports Physiol Perform 9(2):365–370
39. Davis M, Luescher R (2008) Oar forces from unobtrusive optical fibre sensors. Technical report, Australian Institute of Sport
40. Rogers AJ (1988) Distributed optical-fibre sensors for the measurement of pressure, strain and temperature. Phy Rep 169(2):99–143
41. Hull MLA, Mote Jr CDB (1980) Leg loading in snow skiing: computer analyses. J Biomech 13(6):481–491
42. Quinn TP, Mote Jr CD (1991) Optimal design of an uncoupled six-degree-of-freedom dynamometer 30(1):40–48, Mar 1990

43. Yoneyama T, Kitade M, Osada K (2010) Investigation on the ski-snow interaction in a carved turn based on the actual measurement. Procedia Eng 2(2):2901–2906
44. Nakazato K, Scheiber P, Müller E (2013) Comparison between the force application point determined by portable force plate system and the center of pressure determined by pressure insole system during alpine skiing. Sports Eng 16(4):297–307
45. Stricker G, Scheiber P, Lindenhofer E, Müller E (2010) Determination of forces in alpine skiing and snowboarding: validation of a mobile data acquisition system. Eur J Sport Sci 10:31–41
46. Hoset M, Rognstad AB, Rølvåg T, Ettema G, Sandbakk Ø (2013) Construction of an instrumented roller ski and validation of three-dimensional forces in the skating technique. Sports Eng 17(1):1–10
47. Scott N, Yoneyama T, Kagawa H, Osada K (2007) Measurement of ski snow-pressure profiles. Sports Eng 10(3):145–156
48. Kiefmann A, Krinninger M, Lindemann U, Senner V, Spitzenpfeil P (2006) A new six component dynamometer for measuring ground reaction forces in alpine skiing. In: The engineering of sport 6 Springer
49. Senner V, Michel FI, Lehner S (2013) Ski equipment-related measures to reduce knee injuries. International Sports Engineering Association
50. Nilsson J, Karlöf L, Jakobsen V (2013) A new device for measuring ski running surface force and pressure profiles. Sports Eng 16(1):55–59
51. Ohtonen O, Lindinger S, Lemmettylä T, Seppälä S, Linnamo V (2013) Validation of portable 2D force binding systems for cross-country skiing. Sports Eng 16(4):281–296
52. Swarén M, Therell M, Eriksson A, Holmberg HC (2013) Testing method for objective evaluation of cross-country ski poles. Sports Eng 16(4):255–264
53. Pellegrini B, Bortolan L, Schena F (2011) Poling force analysis in diagonal stride at different grades in cross country skiers. Scand J Med Sci Sports 21:589–597
54. Robert-Lachaine X, Turcotte RA, Dixon PC, Pearsall DJ (2012) Impact of hockey skate design on ankle motion and force production. Sports Eng 15(4):197–206
55. Mustone TJ (2000) Characterizing the performance of baseball bats using experimental and finite element methods. In: 3rd International conference on the engineering of sport, June, Sydney, Australia
56. Allen T, Foster L, Carré M, Choppin S (2012) Characterising the impact performance of field hockey sticks. Sports Eng 15(4):221–226
57. Krumm D, Schwanitz S, Odenwald S (2013) Development and reliability quantification of a novel test set-up for measuring footwear bending stiffness. Sports Eng 16(1):13–19
58. McGhie D, Ettema G (2013) Biomechanical analysis of traction at the shoe-surface interface on third-generation artificial turf. Sports Eng 16(2):71–80
59. Guisasola I, James I, Llewellyn C, Stiles V, Dixon S (2009) Quasi-static mechanical behaviour of soils used for natural turf sports surfaces and stud force prediction. Sports Eng 12(2):99–109
60. Murakami M, Seo K, Kondoh M, Iwai Y (2012) Wind tunnel measurement and flow visualisation of soccer ball knuckle effect. Sports Eng 15(1):29–40
61. Chowdhury H, Alam F (2012) Bicycle aerodynamics: an experimental evaluation methodology. Sports Eng 15(2):73–80
62. Ohgi Y, Seo K, Hirai N, Murakami M (2006) Measurement of jumper's body motion in Ski Jumping, In: The engineering of sport 6, Springer, New York, pp 275–280
63. Fuss FK, Niegl G (2008) Instrumented climbing holds and performance analysis in sport climbing. Sports Technol 1(6):301–313
64. Brophy-Williams N, Driller MW, Halson SL, Fell JW, Shing CM (2014) Evaluating the Kikuhime pressure monitor for use with sports compression clothing. Sports Eng doi:10.1007/s12283-013-0125-z
65. Walsh ES, Rousseau P, Hoshizaki TB (2011) The influence of impact location and angle on the dynamic impact response of a hybrid III headform. Sports Eng 13(3):135–143

66. Ouckama R, Pearsall DJ (2011) Evaluation of a flexible force sensor for measurement of helmet foam impact performance. J Biomech 44(5):904–909
67. Ramazanoglu N (2013) Transmission of impact through the electronic body protector in Taekwondo. Int J Appl Sci Technol 3(2)
68. Partridge K, Hayes JP, James DA, Hill C, Gin G, Hahn A (2005) A wireless-sensor scoring and training system for combative sports. In: Smart materials, nano-, and micro-smart systems 5649:402–408. International Society for Optics and Photonics
69. de Leva P (1996) Adjustments to Zatsiorsky-Seluyanov's segment inertia parameters. J Biomech 29(9):1223–1230
70. Lee JC (2008) Hacking the nintendo wii remote. Pervasive Comput 7(3):39–45
71. Zihajehzadeh S, Loh D, Lee TJ, Hoskinson R, Park EJ (2015) A cascaded Kalman filter-based GPS/MEMS-IMU integration for sports applications, measurement 73: 200–210, Sept 2015, ISSN 0263-2241
72. James DA (2006) The application of inertial sensors in elite sports monitoring. In: The engineering of sport 6:289–294. Springer New York
73. Davey N, James DA (2003) Signal analysis of accelerometry data using gravity based modelling. Proc SPIE 5274:362–370
74. Ohta K, Ohgi Y, Kimura H, Hirotsu N (2005) Sports Data, Kyoritsu, Tokyo, Japan
75. James D, Busch A, Ohgi Y (2009) Quantitative assessment of physical activity using inertial sensors. In: Pope N, Kuhn KL, Forster JH (eds) Digital sport for performance enhancement and competitive evolution: intelligent gaming technologies, IGI Global
76. Herren R, Sparti A, Aminian K, Schutz Y (1999) The prediction of speed and incline in outdoor running in humans using accelerometry. Med Sci Sports Exerc 31:1053–1059
77. Williamson R, Andrews BJ (2001) Detecting absolute human knee angle and angular velocity using accelerometers and rate gyroscopes. Med Biol Eng Comput 39(3):294–302
78. Ozaki H, Ohta K, Jinji T (2012) Multi-body power analysis of kicking motion based on a double pendulum. Procedia Eng 34:218–223
79. Mayagoitia R, Nene A, Veltink P (2002) Accelerometer and rate gyroscope measure-ment of kinematics: an inexpensive alternative to optical motion analysis systems. J Biomech 35:537–542
80. Wong TC, Webster JG, Montoye HJ, Washburn R (1981) Portable accelerometer device for measuring human energy expenditure. IEEE Trans Biomed Eng 6:467–471
81. McGrath D, Greene BR, O'Donovan KJ, Caulfield B (2012) Gyroscope-based assessment of temporal gait parameters during treadmill walking and running. Sports Eng 15(4): 207–213
82. Kavanagh JJ, Morrison S, James DA, Barrett R (2006) Reliability of segmental accelerations measured using a new wireless gait analysis system. J Biomech 39 (15):2863–2872
83. Wixted AJ, Thiel DV, Hahn AG, Gore CJ, Pyne DB, James DA (2007) Measurement of energy expenditure in elite athletes using MEMS-based triaxial accelerometers. IEEE Sens 7(4):481–488
84. Wixted AJ, Billing DC, James DA (2010) Validation of trunk mounted inertial sensors for analysing running biomechanics under field conditions, using synchronously collected foot contact data. Sports Eng 12(4):207–212
85. Ohgi Y, Yasumura M, Ichikawa H, Miyaji C (2000) Analysis of stroke technique using acceleration sensor IC in freestyle swimming. The Engineering of Sport. pp 503–512
86. Davey N, Anderson M, James DA (2008) Validation trial of an accelerometer-based sensor platform for swimming. Sports Technol 1(4–5):202–207
87. Le Sage T, Bindel A, Conway PP, Justham LM, Slawson SE, West AA (2011) Embedded programming and real-time signal processing of swimming strokes. Sports Eng 14(1):1–14
88. James DA, Leadbetter RI, Neeli MR, Burkett BJ, Thiel DV, Lee JB (2011) An integrated swimming monitoring system for the biomechanical analysis of swimming strokes. Sports Technol 4(3–4):141–150

89. Wright BV, Stager JM (2013) Quantifying competitive swim training using accelerometer-based activity monitors. Sports Eng 16(3):155–164
90. Hagem RM, O'Keefe SG, Fickenscher T, Thiel DV (2013) Self contained adaptable optical wireless communications system for stroke rate during swimming. IEEE Sens 13(8): 3144–3151
91. Stamm A, James DA, Thiel DV (2013) Velocity profiling using inertial sensors for freestyle swimming. Sports Eng 16(1):1–11
92. Dadashi F, Crettenand F, Millet GP, Aminian K (2012) Front-crawl instantaneous velocity estimation using a wearable inertial measurement unit. Sensors 12(10):12927–12939
93. Ahmadi A, Rowlands DD, James DA (2010) Development of inertial and novel marker-based techniques and analysis for upper arm rotational velocity measurements in tennis. Sports Eng 12(4):179–188
94. Elliott BC, Marshall RN, Noffal GJ (1995) Contributions of upper limb segment rotations during the power serve in tennis. J Appl Biomech 11:433–442
95. Ahmadi A, Rowlands D, James DA (2009) Towards a wearable device for skill assessment and skill acquisition of a tennis player during the first serve. Sports Technol 2(3–4): 129–136
96. Busch A, James DA (2007) Analysis of cricket shots using inertial sensors. The impact of technology on sport II, pp 317–322
97. Wixted A, Portus M, Spratford W, James DA (2011) Detection of throwing in cricket using wearable sensors. Sports Technol 4(3–4):134–140
98. Ohgi, Y, Baba, T, Sakaguchi I (2005) Measurement of deceleration of Golfer's sway and uncock timing in driver swing motion. The impact of technology on sport, pp 349–354
99. James DA, Uroda W, Gibson T (2005) Dynamics of swing: a study of classical Japanese swordsmanship using accelerometers. In Asia-Pacific Congress on Sports Technology, Australasian Sports Technology Alliance
100. Mattes K, Schaffert N (2010) New measuring and on water coaching device for rowing. J Hum Sport Exerc 5(2):226–239
101. Zhang K, Deakin R, Grenfell R, Li Y, Zhang J, Cameron WN, Silcock DM (2004) GNSS for sports–sailing and rowing perspectives. J Global Positioning 3(1–2):280–289
102. Fuss FK (2009) Design of an instrumented bowling ball and its application to performance analysis in tenpin bowling. Sports Technol 2(3–4):97–110
103. King K, Perkins NC, Churchill H, McGinnis R, Doss R, Hickland R (2011) Bowling ball dynamics revealed by miniature wireless MEMS inertial measurement unit. Sports Eng 13(2):95–104
104. Koyanagi R, Ohgi Y (2008) Measurement of the forces on ball in flight using built-in accelerometer. Proc Int Sports Eng Assoc 7(2):1–8
105. McGinnis RS, Perkins NC (2012) A highly miniaturized, wireless inertial measurement unit for characterizing the dynamics of pitched baseballs and softballs. Sensors 12(9): 11933–11945
106. Fuss FK, Lythgo N, Smith RM, Benson AC, Gordon B (2011) Identification of key performance parameters during off-spin bowling with a smart cricket ball. Sports Technol 4(3–4):159–163
107. Koyanagi R, Ohgi Y (2010) Measurement of kinematics of a flying disc using an accelerometer. Procedia Eng 2(2):3411–3416
108. Kelly D, Coughlan GF, Green BS, Caulfield B (2012) Automatic detection of collisions in elite level rugby union using a wearable sensing device. Sports Eng 15(2):81–92
109. Naunheim RS, Standeven J, Richter C, Lewis LM (2000) Comparison of impact data in hockey, football, and soccer. J Trauma-Inj Infect Crit Care 48(5):938–941
110. Bartsch A, Benzel E, Miele V, Morr D, Prakash V (2012) Impact 'fingerprints' and preliminary implications for an 'intelligent mouthguard' head impact dosimeter. Sports Eng 15(2):93–109

111. Higgins M, Halstead PD, Snyder-Mackler L, Barlow D (2007) Measurement of impact acceleration: mouthpiece accelerometer versus helmet accelerometer. J Athletic Training 42(1):5–10
112. Nuzzo JL, Anning JH, Scharfenberg JM (2011) The reliability of three devices used for measuring vertical jump height. J Strength Conditioning Res 25(9):2580–2590
113. McCamley J, Donati M, Grimpampi E, Mazzà C (2012) An enhanced estimate of initial contact and final contact instants of time using lower trunk inertial sensor data. Gait posture 36(2):316–318
114. McMaster DT, Gill ND, Cronin JB, MCGuigan MR (2013) Is wireless accelerometry a viable measurement system for assessing vertical jump performance? Sports Technol 6(2):86–96
115. Harding JW, Small JW, James DA (2007) Feature extraction of performance variables in elite half-pipe snowboarding using body mounted inertial sensors. In: Microelectronics, MEMS, and nanotechnology, International Society for Optics and Photonics, pp 679917–679917
116. Harding JW, Mackintosh CG, Martin DT, Hahn AG, James DA (2006) Automated scoring for elite half-pipe snowboard competition: important sporting development or techno distraction? Sports Technol 1(6):277–290
117. Helten T, Brock H, Müller M, Seidel HP (2011) Classification of trampoline jumps using inertial sensors. Sports Eng 14(2–4):155–164
118. Ohgi Y, Hirai N, Murakami M, Seo K (2008) Aerodynamic study of ski jumping flight based on inertia sensors. In: The engineering of sport 7:157–164. Springer, Paris
119. Supej M (2010) 3D measurements of alpine skiing with an inertial sensor motion capture suit and GNSS RTK system. J Sports Sci 28(7):759–769
120. Gilgien M, Spörri J, Chardonnens J, Kröll J, Müller E (2013) Determination of external forces in alpine skiing using a differential global navigation satellite system. Sensors (Basel) 13(8):9821–9835
121. Rosenbaum D, Hautmann S, Gold M, Claes L (1994) Effects of walking speed on plantar pressure patterns and hindfoot angular motion. Gait posture 2(3):191–197
122. Gullstrand L, Nilsson J (2009) A new method for recording the temporal pattern of stride during treadmill running. Sports Eng 11(4):195–200
123. Türk-Noack U (1994) Field trial of the LAVEG laser diode system for kinematic analysis in various kinds of sports. In: ISBS-conference proceedings archive 1(1)
124. Poirier L, Lozowski EP, Maw S, Stefanyshyn DJ, Thompson RI (2011) Experimental analysis of ice friction in the sport of bobsleigh. Sports Eng 14(2–4):67–72
125. Clark RA, Pua YH, Fortin K, Ritchie C, Webster KE, Denehy L, Bryant AL (2012) Validity of the Microsoft Kinect for assessment of postural control. Gait posture 36(3): 372–377
126. Gokturk SB, Yalcin H, Bamji C (2004) A time-of-flight depth sensor-system description, issues and solutions. In: Computer vision and pattern recognition workshop, CVPRW'04. Conference on IEEE, pp 35–35
127. Petushek E, Richter C, Donovan D, Ebben WP, Watts PB, Jensen RL (2012) Comparison of 2D video and electrogoniometry measurements of knee flexion angle during a countermovement jump and landing task. Sports Eng 15(3):159–166
128. Comstock BA, Solomon-Hill G, Flanagan SD, Earp JE, Luk HY, Dobbins KA, Kraemer WJ (2011) Validity of the Myotest® in measuring force and power production in the squat and bench press. J Strength Conditioning Res 25(8):2293–2297
129. Witters J, Heremans G, Bohets W, Stijnen V, Van Coppenolle H (1985) The design and testing of a wire velocimeter. J Sports Sci 3:197–206
130. Fischer O, Braune W (1895) Der Gang des Menschen: Versuche am unbelasteten und belasteten Menschen, Band 1. (In German). Hirzel Verlag

131. Nordquist J, Hull ML (2007) Design and demonstration of a new instrumented spatial linkage for use in a dynamic environment: application to measurement of ankle rotations during snowboarding. J Biomech Eng 05/2007 129(2):231–239.
132. Petrone N, Marcolin G, Panizzolo FA (2013) The effect of boot stiffness on field and laboratory flexural behavior of alpine ski boots, Sports Eng 12/2013 16(4)
133. Fong DT, Chan YY (2010) The use of wearable inertial motion sensors in human lower limb biomechanics studies: a systematic review. Sensors (Basel) 10(12):11556–11565
134. Oggiano Luca, Brownlie Len, Troynikov Olga (2013) Lars Morten Bardal, Camillan Sæter, Lars Sætran, A Review on Skin Suits and Sport Garment Aerodynamics: Guidelines and State of the Art. Procedia Eng 60:91–98
135. Schlader Zachary J, Simmons Shona E, Stannard Stephen R, Mundel Toby (2011) Skin temperature as a thermal controller of exercise intensity. Eur J Appl Physiol 111:1631–1639
136. Joustra JJ, Jansen AJ (2014) Thermal simulation of a dinghy sailor. Procedia Eng 72: 672–677
137. Hofer Patrick, Hasler Michael, Fauland Gulnara, Bechtold Thomas (2011) Werner Nachbauer temperature, relative humidity and water absorption in ski boots. Procedia Eng 13:44–50
138. Colonna M, Moncalero M, Nicotra M, Pezzoli A, Fabbri E, Bortolan L, Pellegrini B, Schena F (2014) Thermal behaviour of ski-boot liners: effect of materials on thermal comfort in real and simulated skiing conditions. Procedia Eng 72:38–391
139. Bulut J, Janta M, Senner V, Kreuzer J (2013) Determination of insulation properties of functional clothing using core body temperature gradients as quantification parameter, Procedia Eng 60:208–213, ISSN 1877–7058
140. Corbett J, Barwood MJ, Tipton MJ (2014) Physiological cost and thermal envelope: a novel approach to cycle garment evaluation during a representative protocol, Scand J Med Sci Sports, Jan 2014
141. Watson C, Nawaz N, Troynikov O (2013) Design and evaluation of sport garments for cold conditions using human thermoregulation modeling paradigm. Procedia Eng 60:151–156
142. Lee J, Kim S, Lee J, Yang D, Park BC, Ryu S, Park I (2014) A stretchable strain sensor based on a metal nanoparticle thin film for human motion detection, Nanoscale. 1 Sep 2014
143. Yao S, Zhu Y (2014) Wearable multifunctional sensors using printed stretchable conductors made of silver nanowires. Nanoscale 6(4):2345–2352
144. Kuang J, Liu L, Gao Y, Zhou D, Chen Z, Han B, Zhang Z (2013) A hierarchically structured graphene foam and its potential as a large-scale strain-gauge sensor. Nanoscale 5(24):12171–12177
145. Yang T, Wang Y, Li X, Zhang Y, Li X, Wang K et al (2014) Torsion sensors of high sensitivity and wide dynamic range based on a graphene woven structure. Nanoscale 6(21):13053–13059
146. Park S, Vosguerichian M, Bao Z (2013) A review of fabrication and applications of carbon nanotube film-based flexible electronics. Nanoscale 5(5):1727–1752
147. Dugan RE, Gabriel KJ (2013) "Special Forces" innovation: how DARPA attacks problems. Harvard Bus Rev 91(10):74–84
148. Justham L, Slawson S, West A, Conway P, Caine M, Harrison R (2008) Enabling technologies for robust performance monitoring. In: The engineering of sport 7:45–54 Springer, Paris
149. Fleming P, Young C, Dixon S, Carré M (2010) Athlete and coach perceptions of technology needs for evaluating running performance. Sports Eng 13(1):1–18
150. Ringuet-Riot CJ, Hahn A, James DA (2014) A structured approach for technology innovation in sport. Sports Technol 6(3):137–149
151. Van Boeijen A, Daalhuizen J, Zijlstra J, van der Schoor R (eds) (2013) DELFT Design Guide. BIS Publishers, Amsterdam Author: Annemiek van, Jaap, Jelle Zijlstra, Roos van der Schoor

152. Cao L, Ramesh B (2008) Agile requirements engineering practices: an empirical study. IEEE Softw 25(1):60–67
153. D'Amico A, Di Natale C (2001) A contribution on some basic definitions of sensors properties. IEEE Sensors J 1:183–190
154. Giubilato F, Petrone N (2012) A method for evaluating the vibrational response of racing bicycles wheels under road roughness excitation. Procedia Eng 34:409–414
155. Konrad P (2005) The abc of emg. A practical introduction to kinesiological electromyography 1
156. Cutmore T, James DA (1999) Identifying and reducing noise in psychophysiological recordings. Int J Psychophysiol 32:129–150
157. Horowitz P, Hill W, Hayes TC (1989) The art of electronics. Cambridge University Press, Cambridge
158. Analog Devices (2014) ADIS16407 Ten degrees of freedom Inertial Sensor, Datasheet and product info, http://www.analog.com/en/mems-sensors/mems-inertial-measurement-units/adis16407/products/product.html. Accessed 4 Feb 2014
159. James DA, Channells J, MadhusudanRao, N, Thiel DV (2005) An embedded wireless sensor network at 433 MHz for agricultural applications. In: Microelectronics, MEMS, and nanotechnology, International Society for Optics and Photonics, pp 603516–603516
160. Institute of Electrical and Electronics Engineers (IEEE), IEEE 802: overview and archecture. Retrieved from http://standards.ieee.org/about/get/802/802.html. 2 Feb 2014
161. Dynastream Innovations Inc (2014) Welcome to the world of ANT, Retrieved from http://www.thisisant.com/developer/ant/ant-basics/. 15 Feb 2014
162. James DA, Fanella L, Cusani R (2013) Near real time network simulation for team sports monitoring. Procedia Eng 60:422–427
163. http://www.fieldlabs.eu/. Accessed on 15 Sept 2014
164. Woellik H, Mueller A, Herriger J (2014) Permanent RFID timing system in a track and field athletic stadium for training and analysing purposes, Procedia Eng 72:202–207, ISSN 1877-7058, http://dx.doi.org/10.1016/j.proeng.2014.06.034
165. James DA, Wixted A (2011) ADAT: a matlab toolbox for handling time series athlete performance data. Procedia Eng 13:451–456
166. Hagem RM, Haelsig T, Steven G, O'Keefe SG, Stamm A, Fickenscher T, Thiel DV (2013) Second generation swimming feedback device using a wearable data processing system based on underwater visible light communication, Procedia Eng 60:34–39
167. Cesarini D, Schaffert N, Manganiello C, Mattes K (2014) AccrowLive: a multiplatform telemetry and sonification solution for rowing. Procedia Eng 72:273–278
168. Kooyman DJ, James DA, Rowlands DD (2013) A feedback system for the motor learning of skills in golf, Procedia Eng 60:226–231, ISSN 1877-7058
169. Harfield F, Halkon B, Mitchell S, Phillips I, May A (2014) A novel, real-time biomechanical feedback system for use in rowing. Procedia Eng 72:126–131
170. James DA, Simjanovic M, Leadbetter R, Wearing S (2014) Design and test of a custom instrumented leg press for injury and recovery intervention. Procedia Eng 72:38–43
171. Müller M (2010) Enhancing sport—sports technology design in the context of sport motive, motion task and product feature, doctoral dissertation, TUM, Munich
172. Swinnen S (1996) Information feedback for motor skill learning: a review, advances in motor learning and control, Zelaznik HN (ed), Human kinetics
173. Axisa F, Schmitt PM, Gehin C, Delhomme G, McAdams E, Dittmar A (2005) Flexible technologies and smart clothing for citizen medicine, home healthcare, and disease prevention. IEEE Trans Inf Technol Biomed 9(3):325–336
174. Bak D (2003) Rapid prototyping or rapid production? 3D printing processes move industry towards the latter. Assembly Autom 23(4):340–345
175. James DA, Hayes JP, Davey N (2007) From conception to reality: a wearable device for automated swimmer performance analysis. In: Japan society for sciences in swimming and water exercise 11th annual conference. Japan Society for Sciences in Swimming and Water Exercise

176. Schaller RR (1997) Moore's law: past, present and future. IEEE Spectr 34(6):52–59
177. Chan M, Estève D, Fourniols JY, Escriba C, Campo E (2012) Smart wearable systems: current status and future challenges. Artif Intell Med 56(3):137–156
178. Wisbey B, Montgomery PG, Pyne DB, Rattray B (2010) Quantifying movement demands of AFL football using GPS tracking. J Sci Med Sport 13(5):531–536
179. Swan M (2012) Sensor mania! the internet of things, wearable computing, objective metrics, and the quantified self 2.0. J Sens Actuator Netw 1(3):217–253
180. Dugan RE, Gabriel KJ (2013) "Special Forces" innovation: how DARPA attacks problems. Harvard Bus Rev 91(10):74–84
181. Fleming P, Young C, Dixon S, Carré M (2010) Athlete and coach perceptions of technology needs for evaluating running performance. Sports Eng 13(1):1–18
182. Chan M, Estève D, Fourniols JY, Escriba C, Campo E (2012) Smart wearable systems: Current status and future challenges. Artif Intell Med 56(3):137–156
183. Axisa F, Schmitt PM, Gehin C, Delhomme G, McAdams E, Dittmar A (2005) Flexible technologies and smart clothing for citizen medicine, home healthcare, and disease prevention. Inf Technol Biomed IEEE Trans 9(3):325–336
184. Fortunati L (2002) The mobile phone: Towards new categories and social relations 1. Inf Commun Soc 5(4):513–528
185. Someya T (2013) Building bionic skin. IEEE Spectr 50(9):50–56
186. Thiel DV (2008) Sustainable electronics: wireless systems with minimal environmental impact. In: Antennas, propagation and EM theory, 2008. ISAPE 2008. 8th International Symposium on, pp 1298-1301. IEEE, Nov 2008
187. Roco MC (2003) Nanotechnology: convergence with modern biology and medicine. Cur Opin Biotechnol 14(3):337–346
188. KPMG (2103) The SMAC code: embracing new technologies for future businesses. KPMG International http://www.kpmg.com/IN/en/IssuesAndInsights/ArticlesPublications/Documents/The-SMAC-code-Embracing-new-technologies-for-future-business.pdf. Accessed on 28 Apr 2014
189. Lamkin P (2016) Wearable tech market to be worth $34 billion by 2020, Forbes Magazine, http://www.forbes.com/sites/paullamkin/2016/02/17/wearable-tech-market-to-be-worth-34-billion-by-2020/
190. Carlaw S (2013) Emerging bluetooth verticals, Bluetooth World Shanghai
191. McNab T, James DA, Rowlands D (2011) iPhone sensor platforms: applications to sports monitoring. Procedia Eng 13:507–512

